East Asian Social Science Monographs

Malays in Singapore

Malays in Singapore
Culture, Economy, and Ideology

Tania Li

SINGAPORE
OXFORD UNIVERSITY PRESS
OXFORD NEW YORK
1989

Oxford University Press

Oxford New York Toronto
Delhi Bombay Calcutta Madras Karachi
Petaling Jaya Singapore Hong Kong Tokyo
Nairobi Dar es Salaam Cape Town
Melbourne Auckland
and associated companies in
Berlin Ibadan

Oxford is a trade mark of Oxford University Press

© Oxford University Press Pte. Ltd. 1989

Published in the United States by
Oxford University Press, Inc., New York

All rights reserved. No part of this publication may be reproduced,
stored in a retrieval system, or transmitted, in any form or by any means,
electronic, mechanical, photocopying, recording or otherwise,
without the prior permission of Oxford University Press

ISBN 0 19 588914 2

The author and publisher have made every effort to trace
the people depicted in the cover illustration, but without success.
To them, the author and publisher offer their apologies,
trusting that they will accept the will for the deed.

British Library Cataloguing in Publication Data

Li, Tania, 1959-
    Malays in Singapore: culture, economy and
    ideology.—(East Asian social science monographs).
    1. Singapore. Social life
    I. Title. II. Series.
959.5'705
    ISBN 0-19-588914-2

Library of Congress Cataloging-in-Publication Data

Li, Tania, 1959-
    Malays in Singapore: culture, economy, and ideology/Tania Li.
    p. cm.—(East Asian social science monographs)
    Bibliography: p.
    Includes index.
    ISBN 0-19-588914-2:
    1. Malays (Asian people)—Singapore. 2. Singapore—Civilization.
3. Singapore—Economic conditions. I. Title. II. Series.
DS598.S742L5 1989   88-38919
959.57—dc19   CIP

Printed in Malaysia by Peter Chong Printers Sdn. Bhd.
Published by Oxford University Press Pte. Ltd.,
Unit 221, Ubi Avenue 4, Singapore 1440

*To my father, who loved Singapore and all its peoples,
and admired their many achievements.*

Preface

AN account of the context and methods of research is important to the reader of a sociological or anthropological study, since it provides the reader's only direct communication with the researcher, who then, by convention, becomes 'invisible' in the text. Every book is influenced by the theoretical preconceptions and research methods adopted by its author. Studies based extensively on material obtained from ordinary people in the contexts of direct and informal interaction are prone to additional, more personal influences. A short account of the researcher and the research process is therefore in order.

Field research and data gathering for this study were carried out in Singapore between April 1983 and November 1984. After an initial period of language learning, all the work was carried out by the author in Malay. Singapore is an urban society where about 80 per cent of the population live behind the closed doors of high-rise flats. Street-corner or casual observation is therefore not a very productive source of data, particularly on household questions. Various techniques were used in the attempt to achieve familiarity with the everyday life of Singapore Malay families.

Household data were collected by a combination of semi-formal and informal means. From a sample of 70 households, basic data were collected on personal history, subgroup origins, household composition, education, occupation, income, monthly budget, association memberships, and neighbourhood ties. These data were usually obtained from the woman of the house during the day when the husband was absent, as part of a strategy to ensure that women's, as well as men's, perspectives would be obtained. Some households were contacted by means of introductions, but the risk of bias in the selection of households proved too great, so that the bolder approach of knocking directly on doors unannounced was then adopted. The researcher's status as a young, white mother was unthreatening, yet curious enough to the Malay housewives that they were willing to invite her in and answer questions. In return, they often asked the researcher very similar questions about her own family structure and relationships, making the 'interview' more of an exchange.

Following the initial survey-type interview, a further interview was conducted with the husbands of those households selected for further study. A tape recorder was used on this occasion and the interviewee was asked to recount his life story. This technique generated half an hour to three hours of narrative, which was then transcribed, and which provided data on personal and family background, migration, working history, and perceptions of life chances, ambitions, and constraints. Though the researcher did not raise ethnic or political issues directly, these usually entered the account either as general comments or specific examples volunteered by the speaker. Thirty life histories were obtained and transcribed.

A deeper relationship was developed with about 30 of the 70 surveyed households by frequent, informal visiting and participation in various activities, providing opportunities to talk to other family members and to observe interaction inside and outside the household. Since the households were chosen in groups of ten neighbours, it was possible to observe and hear accounts of the relationships between parties all of whom were familiar to the researcher. The neighbourhoods selected, spanned the range of income groups and were in different parts of the city to ensure exposure to households from all backgrounds. Ten households were from a kampong in Pasir Panjang, 40 were from Queenstown in one- and three-room flats, and 20 were from Bedok in two- and five-room flats.

Additional data on the history and origins of Singapore Malays were obtained from interviews with the elders of several kampongs. The data on business were collected by means of taped interviews with 30 entrepreneurs in formally registered companies and 20 owners of informal enterprises. Feature articles on business in the local Malay press provided additional information and discussions were held with members of the Malay Chamber of Commerce. The data on divorce were obtained from a three-week period of observation at the Syariah Court and discussions with some of its officials. Data on the Malay élite were obtained from interviews with Malay Members of Parliament, journalists, teachers, and leaders and members of Malay educational, cultural, and religious organizations. The researcher attended a three-day religious training course for Muslim undergraduates and a three-day seminar on the Malay problem attended by 200 Malay leaders in October 1985.

Four months were spent gathering statistical and other secondary data in the libraries of the National University of Singapore and the Institute of Southeast Asian Studies. Close attention was paid to current events and debates described in the national English and Malay newspapers, and opportunities were taken to discuss these with members of the Malay community in order to assess the relations between Malay and National concerns. Printed materials relating to surveys, campaigns, or seminars within the Malay community over the past two decades were made available to the researcher from the private collections of participants or organizers.

Dalhousie University, TANIA LI
Halifax, Canada
August 1988

Acknowledgements

THIS work is based upon a doctoral dissertation submitted to the Department of Social Anthropology, University of Cambridge, in 1986. Fieldwork was carried out in Singapore from April 1983 until November 1984. The research was financed by the British Economic and Social Research Council, with supplementary funds from King's College, Cambridge, the Evans Trust, and the Smuts Fund. The School for Resource and Environmental Studies at Dalhousie University, my current professional home, provided funds for the preparation of the manuscript for publication.

Thanks are due to my doctoral supervisors, Alan Macfarlane and the late Barbara Ward, for their interest in my work and their whole-hearted support of this project. Members of the Singapore academic community who helped me in practical ways and by taking time to read and discuss my early research findings were John Clammer, Sharon Siddique, Vivienne Wee, Normala Manap, Suriani Suratnam, Mariam Ali, and Pang Keng Fong. The thesis examiners were William Wilder and Leo Howe, both of whom provided useful criticism which guided the process of revision for publication.

My husband Victor Li, mother Anita, parents-in-law, and various baby-sitters deserve many thanks for their support and encouragement and for helping me to take care of one, two, then three children who came into existence during the same period as this book. Victor also provided valuable editorial advice and maintained a sceptical and therefore stimulating perspective on anthropology in general, and my work in particular.

My greatest debt is to the Singapore Malay families, business people, and community leaders who gave generously of their time and knowledge to contribute to this research. Most spoke to me in confidence and asked not to be named, either because of the personal nature of the information shared with me or because of their concern that research on the Malay community, whatever its topic or intent, might be deemed 'sensitive'. It is my sincere hope that this work, which brings an outsider's perspective to cultural, economic, and ideological processes, may clarify some issues and contribute in a positive manner to discussions within the Malay community about their past and future. I am confident that both Malay and other readers will be able to extract and interpret the parts they find useful in the light of their own pre-understandings and concerns, just as I, as an observer, was engaged in an interpretive act while looking on and trying to make sense of a particular form of life.

Note

All dollar figures in the text and accompanying tables are Singapore dollars, which, unless otherwise specified, refer to 1982-3 values.

Contents

Preface	vii
Acknowledgements	ix
Tables	xiv
Introduction	xv

PART I–CULTURE, ECONOMY, AND THE HOUSEHOLD — 1

1 Conceptual Framework: Householding and Malay Kinship — 3
  Householding — 4
  Kinship, Commodification, and the Gift — 6

2 Household Membership and Consumption Rights — 13
  Household Membership — 13
  Consumption Rights — 15

3 Householding: Husband and Wife — 18
  The Obligations of Maintenance and Housework — 18
  Budgetary Margins and the Distribution of Rewards — 20
  Domestic Labour: An Undervalued Commodity — 23
  Property Division: Separate Economies and the Balance of Power — 25
  Householding: The Malay Cultural Heritage and Islamic Law — 29
  Divorce — 33

4 Householding: Parents and Children — 41
  The Care of a Young Child — 43
  Parents and Working Children — 44
  Parents and Unmarried Working Daughters — 49
  Parents and Unmarried Working Sons — 53
  Authority and the Flow of Giving — 56
  Parents and Children: Old Age — 60
  The Personal Life Span and the Afterlife — 67
  Kinship Sentiment and the Value of Children — 72

5 Householding Relationships and the Expansion of Economic Resources 75
Individual Salvation and the Dispersal of Inheritance 76
Parental Investment in Children: Land and Education 78
The Malay Household and Business Enterprise 83

PART II–STRUCTURING PRACTICES: CULTURAL AND ECONOMIC DIFFERENTIATION IN SINGAPORE 1959-1984 89

Introduction to Part II 91

6 The Formation of the Singapore Malay Community 93
The Pattern of Migration and Settlement 93
Ethnic Sentiment and the Development of New Cultural Forms 97

7 Malays in the National Economic and Education System 99
The Economic Position of Malays and Chinese Prior to 1959 100
Chinese and Malay Economic Positions 1959-1980 102
Labour-force Participation and Household Size 104
Entrepreneurship 106
Discrimination 108
The Distribution of Opportunities and Rewards in the National Economy 112
The National Education System 114
Malays and Education 116

8 The Cultural Basis of Class Differentiation 122
The Impact of External Conditions on Household Form 123
The Sense of Class and the Achievement of Social Mobility 124
Individual and Community: The Cultural Heritage 129
Class and Hierarchy in the Presence of the Chinese 133

9 Malay Entrepreneurship 137
The Malay Niche in Entrepreneurship 138
Entrepreneurship within the Malay Community 141

10 The Competitive Context: Chinese Householding and Entrepreneurship 149
The Chinese Household 149
The Chinese Family and Business 158

PART III–CULTURE, ECONOMY, AND IDEOLOGY — 163

Introduction to Part III — 165

11 Culture, Economy, and Ideology — 166
Culture and Malay Backwardness: The Development of the Orthodox View — 168
Economic Differentiation and Cultural Orthodoxy in the 1980s — 173
Ideology in Singapore: Cultural Explanations for Structural Inequalities — 178

12 Conclusion — 184

Glossary — 187
Bibliography — 189
Index — 201

Tables

2.1	Malay Household Membership, 1957 and 1980	13
4.1	Malay Plan of Living Arrangements in Old Age	62
4.2	Malay Perceived Sources of Important Financial Support in Old Age	63
7.1	Malay and Chinese Monthly Household Incomes and Percentage of Households in Poverty, 1953	100
7.2	Educational Level of Malay and Chinese Male Heads of Household, 1953	101
7.3	Malay and Chinese Working Males Aged 10 Years and Over by Occupation, 1957, 1970, and 1980	102
7.4	Employed Male Malays and Chinese by Monthly Income, 1975, 1978, and 1980	103
7.5	Malay and Chinese Female Economic Activity Rate by Age, 1957, 1970, and 1980	104
7.6	Malay and Chinese Working Females Aged 10 Years and Over by Occupation, 1980	105
7.7	Malay and Chinese Total Fertility Rate, 1957-1976	105
7.8	Malay and Chinese Females who have ever been Married, by Highest Qualification and Mean Number of Children Born Alive	106
7.9	Malay and Chinese Working Males Aged 10 Years and Over by Employment Status, 1957, 1970, and 1980	106
7.10	Income of Males Aged 10 Years and Over by Education and Employment Status, 1980	107
7.11	Number of Establishments (Commerce and Manufacture) by Size (Number of Workers), 1970 and 1980	108
7.12	Estimates of Poverty, 1954 and 1973	112
7.13	Highest Qualification Attained by Malay and Chinese Males by Age Cohort, 1980	117

Introduction

SINCE this work is not a conventional ethnography, and it does not attempt to describe the totality of Singapore Malay life, it is as well to begin with a brief discussion of the theoretical framework which provided the focus for data gathering and which informs the analysis. The central question running through this book is this: how do the material conditions of life at the national level affect the day-to-day social practices of individuals and groups, and how do the culturally informed activities of daily life, through their cumulative and unintended impact, shape broad national economic and social conditions? No abstract answer is attempted here. Rather, the 'how' is answered by describing, exploring, and examining the nature of the interaction between macro-level economic and institutional conditions, inherited cultural forms and ideas, and the observable micro-level of day-to-day practices in the household and the community.

The study analyses the ways in which the Malay cultural heritage, and general economic conditions in contemporary Singapore, combine and interact in shaping the form of Malay household and community life. The study focuses on the creative ways in which cultural ideas are adapted to meet new conditions, and on the way that culturally informed practices, in turn, shape the conditions of daily life for individuals and contribute to social and economic processes at the national level.

The account of Singapore Malay life provided here acknowledges the role of both cultural ideas and macro-economic factors in shaping day-to-day practices, while not regarding either economic or cultural factors as static or pre-given. Cultural ideas are shown to be continuously created, recreated, and adapted in the course of daily life, in the light of new knowledge and new circumstances. Features of the economy, such as the availability of work for individuals of different age and sex, and rates of pay, form the practical conditions of daily life, and yet they do not have a mechanistic or determining effect upon social life because they are perceived, interpreted, and acted upon in the light of cultural knowledge.

The idea that social life is creative, and yet also constrained by the cultural and material order that pre-exists the individual, is expressed in Marx's well-known statement that 'Men make their own history, but they do not make it just as they please; they do not make it under circumstances chosen by themselves, but under circumstances directly encountered, given and transmitted from the past' (Marx, 1951: 225). This idea has been developed by Giddens (1979), Bourdieu (1977), and others, and is well summarized by Giddens (1985: 170):

> All social life is inherently creative, in the sense that it is carried in and through the knowledgeable activities engaged in by agents in the course of their day to day lives. Social systems do not have form, or continuity across time and space, because they are programmed to do so by some mysterious set of causal forces propelling individuals along pre-prepared pathways. On the other hand, situated actors do not create most of what they do in the sense of inventing it ex nihilo. Quite to the contrary, they act in social contexts whose modes of organization precede their existence in time, and spread out laterally in space.

The term 'culture' is used in this work to refer to the spoken or tacit knowledge of human agents which is the source, medium, and outcome of daily activities. In accepting the premises that individuals 'act in social contexts whose modes of organization precede their existence in time, and spread out laterally in space', it becomes necessary to go beyond the (mythical) ethnographic present of Singapore Malay life in the 1980s to obtain a broader perspective on the Malay cultural heritage. Comparative ethnographic data from other times and places in the Malay world are used to provide the necessary historical and geographical perspective on contemporary cultural forms.

Cultural variation across time and space is a normal feature of social life in all contexts, rural and urban, when social life is viewed in terms of creative practices. The use of comparative data is not intended to imply that practices in one location, such as rural Malaysia or Java, are the pristine or original source from which urban Singapore Malay culture is derived. Rather, the benefit of comparative ethnographic data is in helping to clarify the depth of cultural continuity. Where social practices observed in contemporary Singapore are found to be similar to those elsewhere in the Malay world, despite different economic and institutional conditions, these practices can be said to be deeply embedded in the Malay cultural heritage (though never static). Where variations in practice occur across time and space, the analyst receives a signal that different structural, economic, or other conditions are playing a critical role. In this work, comparative ethnographic data are used extensively where they provide interpretive insight, but they are not intended to provide simple or deterministic explanations. Cultural knowledge suggests ways of acting, but it does not determine the path of social life which reflects unique sets of economic and institutional conditions.

A special opportunity for understanding Malay social life in Singapore is presented by the presence of other ethnic groups. These groups draw on their own distinct cultural heritages (with their own historical and geographical parameters) while acting within economic and institutional contexts that are often the same for all Singapore citizens. Comparison of the practices of Malays with those of other groups, such as the Chinese, provides a means with which to assess the relative importance of cultural factors versus structural conditions in the interpretation of a particular social form. Such a comparison does not rely on a simplistic dichotomization of one static culture (the Chinese) versus another (the Malays); rather, the comparison of different forms of life provides a source of insight contributing to the process of analysis and interpretation.

The 'social contexts whose modes of organization precede [the individual's] existence in time, and spread out laterally in space' are economic and political as well as cultural. Published data, such as national censuses and statistics, educational and economic policies and statements, are used in this study to obtain an understanding of the macro-level contexts in which Singapore Malay life is pursued. The analysis shows how some important social, economic, and institutional features of the national order are themselves created and sustained as the cumulative outcome of day-to-day practices, albeit often unintended and unknown to individual actors. National statistical data are particularly useful in this regard, as they tend to reveal cumulative trends not always obvious to either participants or casual observers. While statistical data are very useful in complementing ethnographic observation, greatly increasing the power and scope of the analysis, the reverse is also true. If one accepts the premiss that national-level economic and institutional trends are socially created, then it is clear that they cannot be explained or understood without reference to an interpretive, cultural dimension.

The three sections of this work each focus on a different aspect of the interaction between day-to-day social practices, the Malay cultural heritage, and features of the national social, economic, and institutional order. The influence of cultural ideas and economic conditions in shaping the form of Singapore Malay life and, in particular, the Malay household, is examined in Part I. Part II analyses the ways in which cultural practices at the level of daily life in the Malay household and the community have shaped Malay participation in the social and economic development of Singapore since 1959. It looks in particular at the social practices which have contributed to the development of ethnic and class divisions, and to the decline of the socio-economic position of the Malays relative to the Singapore Chinese. Part III examines the origins of the tendency, common in the Malay and national leadership, academic, and popular circles, to interpret Malay economic performance in terms of static cultural stereotypes, without considering economic and ethnic factors or the process of historical change. In contrast to this tendency, the account of Singapore Malay society presented here takes as its central focus the mutual interaction and continuous transformation and change of both cultural and economic forms.

Part I
Culture, Economy, and the Household

1 Conceptual Framework: Householding and Malay Kinship

THE household is chosen here as a focus for ethnographic description and analysis because of its central position in processes of cultural and economic reproduction and change. The household is the primary site of the creation and reproduction of an important subset of cultural knowledge and practices (kinship) and it defines a locale and set of relations within which changes in knowledge and practice and the influence of economic, institutional, and other factors can be observed. The household plays a role in ensuring both cultural and physical continuity from one generation to the next, through the processes of conjugal and intergenerational exchange, and it is thus intricately linked to questions of historical change. Finally, the household, through its routine practices and conscious strategies of its members, links individuals to national processes by preparing them to take up particular roles within a differentiated social, economic, and institutional order.

The household as an institution has received much attention in recent sociological debates. Interest has focused on the relationship between the form of the household and the economic structure of society. The sociological orthodoxy current until the 1970s saw the Western elementary family as the product of the development of industrial capitalism. According to this view, during the process of capitalist development, the family lost its productive functions, developed a strict division of labour between non-productive women and fully waged men, and became elementary in form (Thadani, 1978: 7; Household Research Working Group, 1982: 12). A challenge to this interpretation of the effect of capitalism on the family, arose from research on underdevelopment and the articulation of modes of production, from women's studies, from family historians, and from new specialists in household studies.[1] From this challenge, a new consensus has emerged which argues that there is no direct or simple relation between the economy and the form of the family, since the latter is affected by political and cultural factors specific to its own history. Diverse forms of family persist throughout the process of modernization, and traditional forms can be reinforced in industrial conditions as families adapt to new circumstances. There is not one type of household that is unique or necessary to industrial production, capitalism, or urban life. There are different possible relationships between men, women, young, old, waged, and unwaged, within the household, that enable labour to be sustained and replenished across the generations. Finally, women's domestic work, though often unwaged, is not unproductive since it provides services essential to the formation and maintenance of the formal labour force.

While the new consensus has confirmed the inadequacy of mechanistic interpretations of the effect of the economy on social institutions such as the family, the nature of the relationship between economic and cultural forms still remains unclear. The analysis of the Malay household presented here contributes to this area of debate. The argument moves away from the issue of whether a particular form of family, or a particular form of economy, arose *first*, since such a debate is limited by a reified view of both culture and economy. A cultural form, such as an arrangement of household relationships, does not emerge and cannot be sustained or reproduced in abstraction from specific historical and economic conditions. As Willis (1977: 183) notes in his study of the cultural reproduction of class in Britain, 'cultural activities and attitudes are developed in precise conjunction with real exigencies, and are produced and reproduced in each generation for its own good reasons'.[2] Details on economic conditions in Singapore are reserved for Part II: here it is sufficient to note that the 'real exigencies' faced by the great majority of Singapore Malay men and women are those of the lower-income urban wage-worker. It is the impact of these conditions upon the Malay household that is explored here.

Householding

It is to the inner workings of the household, and the concepts that will be used to analyse it, that the discussion now proceeds. The approach adopted here is that of the Household Research Working Group (1982) which focuses on the process of combination of different labour practices, such as wage work, child care, and domestic work, that is necessary to sustain labour and replenish it across the generations. For this active process of combination of practices, the term 'householding' has been coined. Attention is directed towards the cultural terms in which householding is accomplished, and the variations in this according to changing economic conditions (such as wage rates, job availability by age and sex, and pension provisions). Participants in a householding relationship need not necessarily be co-resident, and for each party the nature and extent of their contribution has to be specified.

The conceptualization of householding as the active process of combination of practices which ensures the current maintenance and generational replenishment of labour, does not assume that these two functions are always congruent or 'naturally' welded together. The interest of the presently productive household members in expanding their current production, or withholding from current consumption, in order to reproduce future generations or maintain elderly dependants, is not guaranteed by their coalescence in a current householding relationship. The cultural and economic basis of transfers between generations must be explicitly analysed, and for this purpose three subcategories of householding practices are identified. 'Current reproduction' refers to those practices that maintain the current consumption standards of the householding group; 'extended reproduction' is the transfers across the generations that sustain the old and raise the young to maturity; 'expanded reproduction' refers to the practices which increase the productive base of the household beyond that needed for current or extended reproduction, and thereby improve the living standards of the current householders or future generations.[3] The extent to which the householding relationship allows for, or encourages, expanded reproduction has significant implications for competitiveness in a differentiated economic order. It will be argued that Malay and Chinese households differ on this crucial point.

Various household members may have different interests in the current, extended, or expanded reproduction of the household. It is necessary to stress this point, because household studies have often been associated with the opposite assumption. The researchers involved with the Household Research Working Group, whose definition of householding is adopted here, make assumptions about the congruence of interests within the household which contradict their claim to explicitly analyse the terms on which householding takes place. They tend to prejudge the issue by asserting that householding is based on the 'pooling of resources' and the 'obligations to share... income' (Household Research Working Group, 1982: 22); 'strategies of coping with the allocation of labour' (Kamal Salih et al., 1983: 20), or 'the household as a decision making unit' (Young and Kamal Salih, 1984: 10). These terms carry a normative weight which assumes either an authority structure that enables the household head to plan, allocate, and form strategies, or a co-operative structure of sharing and pooling. Contrary to this assumption, it will be argued here that though by definition householding accomplishes the reproduction of labour, it may well do so on an *ad hoc* or conflict-ridden basis.

The strongest challenge to the assumption that the household is an income-pooling unit has come from women's studies. Feminist analysis has refuted the thesis that the household is a non-exploitative, intrinsically democratic and co-operative unit operating in the interests of all its members (Whitehead, 1981). Hartman (1981) argues that, far from a unity of interests, the household is a locus of struggle where members have conflicting interests deriving from their relations to production and distribution which are usually structured on an age and gender basis. In so far as it acts as an entity with unified interests, it does so only because of its members' mutual dependence. It is the nature of this dependence and the terms of transfer of goods and services that will be analysed by means of the householding concept.

The assumption common to peasant studies that the peasant household is a unit of production and consumption has also been challenged.

This theory of the peasant household was developed by Chayanov (1966) in the concept of the 'family labour farm', and by Sahlins (1974) in the idea of the 'domestic mode of production'. Diana Wong (1983: 32) criticizes this view and argues that the existence of the family labour farm cannot be assumed, since 'all labour, including family labour, is not merely a value which is "naturally" at the disposal of a household, which is "naturally" formed in the course of an individual's family life cycle, but... it is a social product the formation and control of which has also to be analysed'. Harris (1981) notes that in some societies, men and women own property separately and control its produce; economic transactions between husband and wife can take the form of commodity exchange; and children and their labour cannot be assumed to be under the direct, exclusive control of the household head. Malay society, in both rural 'peasant' and urban industrial conditions, is of this type. Thus there are ethnographic as well as theoretical objections to the assumption that the household is either a corporate or a co-operative unit.

South-East Asian societies have attracted scholarly attention because they are rural based, yet have individualistic patterns of household operation more commonly associated with the modern industrial West or with some tribal societies (Macfarlane, 1978). Insistence on seeing South-East Asians, *qua* rural producers, as necessarily having a 'peasant' or family labour-based system of production has been a source of confusion to at least one writer. McGee (1973) argues that urban petty commodity production in Asia can best be understood as a 'peasant' form of production, its characteristics determined by the tendency of peasants to utilize family labour. This assumption produces what for him is the 'ingenious paradox' of 'peasants in the cities'. He finds confirmation that Chinese 'peasants in the city' do organize their business enterprises on a family basis but he is 'confused' by the evidence that the Javanese do not. He notes the findings of Clifford Geertz that among Javanese bazaar traders, husbands, wives, sons, and siblings all operate independently. He claims that this 'does not exclude per se the possibility of the appropriation of the individual earnings by the head of the household'. He fails to note the evidence of Clifford Geertz (1963) and Dewey (1962) that Javanese parents and children do not pool their earnings and that husbands and wives do so only in a limited way.[4]

The individualism of South-East Asian households makes the question of *what* holds them together and *how* they accomplish householding on a current and extended basis, all the more fascinating. Given the assumption that culture is continuously produced and reproduced in daily life, it will be necessary to look both at the cultural heritage and at the present conditions in Singapore in order to understand the contemporary basis of householding.

Kinship, Commodification, and the Gift

To guide the reader through the detailed analysis to follow, it is useful to present here some Malay kinship ideas, and to introduce in general terms the writer's interpretation of the cultural basis underlying contemporary householding practices in Singapore.

The historical background of the Singapore Malay community will be described in detail in Chapter 6. The key fact to note at this point is that the population which now calls itself 'Malay', is composed of people of varied South-East Asian origins, the biggest groups being from the Malay Peninsula and from Java. For this reason, reference is made to both Malay and Javanese ethnographies to provide general insight into Malay cultural ideas, and these sources indicate that on the question of the basic structure of the household, there are no fundamental differences between Malay and Javanese traditions. There are variations in practices between Malaysia and Java, between rural and urban Malay societies, and across time, and where appropriate, these variations are investigated as they provide a good source of information on the process of cultural change.

Malays are Muslims, and Islam is ingrained in Malay culture. Though some writers have seen Islam as a veneer on pre-Islamic Malay beliefs and practices, this view has to be rejected since it assumes the existence of a primordial, ahistorical, reified Malay culture. The position adopted here concurs with that of Banks (1983: 53) in taking the view 'that regards all of the Malay historical tradition as real, changing, and therefore significant. In this view, Islam has had an important formative and reformative role in molding modern Malay ideas.' Nevertheless, there will be occasions to point out that some individuals are more learned in Islamic canons than others, and make more concerted and explicit attempts to model their behaviour on Islamic principles.

The excellent work by Banks (1983) identifies three sets of kinship ideas which are deeply embedded in Malay culture. One is *darah*, blood, which involves obligations channelled along consanguinal lines, in particular the duties of parents as defined by Islam. The second is *muafakat*, which means agreement or settlement by discussion, and this concerns the calculated, interest-serving aspect of all social relationships, including those within the household and especially between husband and wife. The third idea is *kesayangan*, kinship sentiment or love, which generally (though not necessarily) runs concurrently with *darah* or develops through co-residence, but can be present 'in all close social relationships containing a voluntary moral component' (Banks, 1983: 128). Banks (1983: 48) writes:

This spiritual kinship links the essences of individuality in persons as whole beings. The overt expression of this pure kinship of the spirit is much like The Gift, as explicated by Mauss (1966).... The Malay conception of spiritual kinship expresses the hope that people can live together despite the many problems that beset the attempt. It is the goal that Malay social life seeks to attain so that all Malays will live in a world of affection and esteem.

The value of Banks' analysis is that he has shown how these two different ideas can coexist in one kinship system and even in one relationship. For example, he considers it to be parental duty, defined by *darah* and codified in Islam, to establish the new generation with the economic and social means to lead independent adult lives. Yet, he also finds (1983: 157) that the provisioning of the young with the means of livelihood is seen as a gift, the 'supreme act of kinship', and that 'In dividing up his land before he dies, a man is divesting himself of his worldly property voluntarily in accordance with the unselfish wishes of his heart, and this is viewed as praiseworthy in religious law and public morality ... Malay inheritance is based upon gift or *hadiah*' (Banks, 1983: 138).

It will be argued here that these kinship ideas are intimately linked to an individualistic system of property ownership. Although others have mentioned the individualism of Malays (Djamour, 1959: 46), there has been no systematic analysis of this question in Malay ethnographies, especially as it pertains to relations within the household. In Malay and Javanese society, partners in marriage retain ownership of their own goods and earnings. Parents have complete ownership of their property and during their lifetime may give it to whom they please, not necessarily to their own children. Children, in return, do not have a clearly defined obligation to support their parents. They have no obligation to contribute to the support or material provisioning of their siblings with whom there is no joint estate. In a system based on the idea of individual rights to property, transfers are ultimately based on 'voluntary acts', not legal joint ownership (Macfarlane, 1986).

When the obligations of consanguinity have to be voiced and backed by Islamic law, it is clear that they are not taken for granted or embedded in the property system as such, as would be the case in societies with a corporate family economy, binding elementary family members and possibly extended kin into an indivisible economic unit during at least some stages of the family life cycle. In Malay society, negotiations of self-interest are always a necessary element in joining two or more individuals in a householding relation, since it is not a relation that stems inevitably from any corporate rights. Ultimately, transfers are based on voluntary acts, and this is conceptualized as kinship sentiment, which is a positive exertion of will in the desire to give up the full enjoyment of one's own property and labour, which is an individual right, in order to make gifts and enter into exchanges with others. Kinship sentiment is complementary to ideas of duty and self-interest. It is the voluntary element of will that gives the individual a sense of pride and worthiness as he or she fulfils personal obligations. It is only the willingness of individuals to give and be generous that makes mutual agreement by negotiation possible. This makes the idiom of the gift very prominent in daily life, as it describes both the actual process of negotiation between individuals, and the ideal state of relations. It is, in the Malay view, more worthy to give with love and generosity than to be forced by obligation, and it is more worthy to give than to specify the exact return expected. The best relationships are built on kinship sentiment.

Ethnographic observation established that the idiom of the gift, used to describe and express householding relations, has attained a marked prominence in contemporary Singapore. It is useful to consider here, in more detail, some of the features of the gift as a mode of transferring valued goods and services. Giving, whether of affection and concern or of a material gift, is predicated on the basis of the giver's ownership. A gift is a gift only if it unambiguously belongs to the giver, and the decision to transfer it is a voluntary act of will. So, too, with affection. A sentiment counts as affection only if it is voluntarily given, an idea based on the wholeness and integrity of the donor. That the idiom of the gift should attain prominence *within* the Malay household, rather than between groups or between households as is more common in ethnographic literature of other areas of the world, is related to the integrity of the individual in Malay culture.

The idea of the gift does not imply non-calculation. To be a gift, the object must have a value to both giver and recipient. The 'pure gift', as Mauss (1966) and Bourdieu (1977) explain, is unrealizable because the trace of individuality in the gift sets up a relationship of debt and obligation with the recipient. It is only by uniqueness in measure and timing that the gift is 'carried off' and made into an irreversible transaction. The defining feature of a gift is that exact measure cannot be repaid. The debt thus created poses a threat to the autonomy of the recipient, whose counter-gift becomes not quite the voluntary gift of an individual will, but in a sense a forced response. This tension between the voluntary, willed nature of the gift and its potential to set up relations of obligation and dependence is a permanent feature of the gift. It makes it a powerful vehicle for the creation and maintenance of social relations.

Bourdieu has called the relationship set up by the gift 'symbolic violence'. In the rural economy of Kabylia, he finds (1977: 191) there are

... only two ways (and they prove in the end to be just one way) of getting and keeping a lasting hold over someone: gifts or debts, the overtly economic obligations of debt, or the 'moral', 'affective' obligations created and maintained by exchange.... The reason for the pre-capitalist economy's great need for symbolic violence is that the only way in which relations of domination can be set up, maintained, or restored, is through strategies which, being expressly oriented towards the establishment of relations of personal dependence, must be disguised and transfigured lest they destroy themselves by revealing their true nature; in a word, they must be euphemized.

Seen in this light, it is no surprise that the gift idiom finds currency in the heart of the household. Face-to-face relations in the household could not be more intense, yet it is also the site of the most significant material transactions between people in positions of power and dependence, who must strive to sustain a relationship over a lifetime. This point is recognized by Sahlins (1974: 194) in his discussion of generalized reciprocity:

... for its logical value, one might think of the suckling of children in this context—the expectation of a direct material return is unseemly. At best it is implicit. The material side of the transaction is repressed by the social: reckoning of debts outstanding cannot be overt and is typically left out of account. This is not to say that handing over things in such form, even to 'loved ones', generates no counter obligation. But the counter is not stipulated by time, quantity, or quality: the expectation of reciprocity is indefinite.

When personalized goods and services such as the provision of a home and the care of a child are the elements exchanged, the obligations created are indefinite, and the relationship developed through these gifts is consequently more durable.

The use of the gift idiom is not necessarily a manipulative strategy, nor a form of conspiratorially induced false consciousness as Bourdieu's (1977) account sometimes seems to suggest. The idea of the gift is deeply embedded in the organization of Malay kinship, but it is not an idea imposed or accepted uncritically, nor one that reproduces itself through a dynamic of its own. Individual Malays are able to give an account of the reasons why they seek to organize their household relations on the basis of the gift. It will be shown that they engage in some conscious strategies to try to preserve this element in their kinship relations, and in the course of this attempt, they are creating new cultural practices rather than simply repeating old ones. The reason for the emphasis on gifts is that gifts are thought to build and demonstrate kinship sentiment, which is highly valued in itself, and is also a source of self- and social-esteem. At the same time, it is recognized that gifts build up obligations, so that the element of self-interest is known to be integral to the idea of the gift, even though self-interest may be explicitly denied. To the extent that transactions within the household are seen as gifts, this has a real material effect on the way people proceed in daily life. The choice of idiom is not merely a superficial gloss.

In the analysis to follow, it will be argued that the prominence of the gift idiom in householding relationships in contemporary Singapore is paradoxically related to its conceptual opposite: commodification reaching into the heart of the household. The term 'commodification' refers to the process central to capitalism in which material products and labour power become commodities, seen and measured in terms of money, which acts as a common standard of value (Giddens, 1981: 8). In Singapore the labour power–and hence the very time–of every individual has a value, as there is full employment and waged work is available for everyone. The commercial value of domestic services, such as cooking and child care, is well known. These economic conditions give the housewife a heightened sense of the value of her domestic services and her labour time, and she feels strongly that she is making a gift when she forgoes work opportunities for herself and performs services without pay in the household. The individual's ability to dispose of his or her own time, and its commodified equivalent, money, reinforces the idea that money, goods, and services transferred within the household are *given* by the will of individual owners out of concern and affection for other household members. There is also the recognition, sometimes made explicit, that the personalized nature of household services actually makes them resistant to ultimate commodification. There is no price for the provision of a home or the loving care of a child. The incommensurability of the elements exchanged in the household is retained and this, as noted above, is essential to the idea of the gift.

The assertions most frequently heard when Singapore Malays are discussing transfers of money, goods, and services within the household are that it is not good, or it is troublesome, to calculate things too much. The individual claims to refuse to calculate, and refuses to demand or expect a full recompense, and in this way claims to be making a gift. This gift claim depends on the recognition of imbalances in the exchange of commodities, that is, on the covert and unspoken calculation of values and equivalences. The paradox is that while it is asserted that pure kinship, or the gift, both is and should be the basis of exchange in the household, the calculation of values is integral to the idea of the gift in a commodity economy.

Parry's work on the gift (1986) supports the argument made here on three points: first, in Parry's (1986: 456) concurrence with Mauss that in many, or most, instances there is 'a *combination* of interest and disinterest of freedom and constraint in the gift'; secondly, in the recognition that it is market economic relations (commodification) that engender an emphasis on the pure gift as their conceptual counterpart; and finally, in the observation that the notion of reciprocity in exchange has been overused by analysts imbued with market thinking, since even when all possible types of goods (prestige, emotional well-being, land, and honour, among others) are taken into account, it is not obvious–either to an observer or a participant–that they *must* somehow balance each other.

The capacity of the gift to assert and proclaim the power of the donor is discussed by Bourdieu (1977), and it finds expression in the Malay concept of *kasihan*. This concept is frequently used to describe the transfer of goods and services within the Singapore Malay household. The term is translated as 'kindness, favour; pity; an unfortunate thing' (Coope, 1976). *Kasih*, the root of *kasihan*, means love and affection, and *kasihan* is also related to the word *kasi*, 'to give'. In the way that the term *kasihan* is used to describe transfers of resources within the Malay household, the three senses, pity, love, and the gift, are closely linked. Other students of South-East Asia with whom these research findings were discussed were surprised at the use of this word within the household in Singapore, since it is elsewhere reserved for the treatment of unfortunate outsiders such as beggars. Of all the different terms for gifts in Malay, *kasihan* is the least egalitarian, having strongly condescending implications. Tracing out the pattern of use of this concept will provide some insight into the shifting power relations within the household. It indicates which individuals feel they are giving more than they are receiving, or receiving less than is their real individual entitlement, and hence bestowing the greater gift.

The layering of the concepts of duty, self-interest, and kinship sentiment and the shifts in emphasis and meaning that develop in the context of commodification and in conjunction with the changing economic circumstances of individual household members, male, female, young, and old, will be traced out in detail in the chapters to follow.

1. See Harevin (1977); Sussman and Burchinal (1962); Harris and Young (1981); Stivens (1981); Household Research Working Group (1982); Hartman (1981); Kamal Salih *et al.* (1983); Wallerstein *et al.* (1982); Thadani (1978); and Whitehead (1981).

2. Note that Willis has been criticized for a relapse into material 'determinism in the last instance'. See Jenkins (1983). Nevertheless, his work represents an important attempt to understand the interplay of cultural and material factors by means of a combination of detailed ethnography and examination of broader cultural, economic, and ideological processes. His work was something of a model for the analysis attempted here.

3. Diana Wong (1983) was the source of inspiration for making these distinctions, though the usage developed here is not identical to hers.

4. In a later paper (1982), McGee mentions the individual consumption goals which motivate young Malay women to join the urban work-force, and correctly states that the extent to which there is a family economic strategy and pooling of incomes must be examined rather than assumed.

2 Household Membership and Consumption Rights

Household Membership

IN the Singapore census, a shared hearth is the criterion of a household unit. The census uses the concept of family nucleus to indicate a grouping based on a marital and/or a filial bond, and records households (defined by a shared hearth) which consist of more than one family nucleus. Since the housing unit is not used as a point of reference, the census does not indicate the frequency with which two or more households co-reside in one housing unit, but the practical conditions of living in high-rise flats make this arrangement rare. According to the Singapore census data, Malay households are predominantly elementary in form. They are composed of a single family nucleus living separately from other kin. This pattern changed little during the period 1957-80.

The Malay concept of a household combines the idea of a marital bond with that of the hearth or housing unit. The phrase *berumahtangga* means to marry and to set up home. Upon marriage, the most common residence pattern in Singapore and elsewhere in the Malay world is a short period of residence with either, or periodically both, sets of parents, then separate residence as an elementary family within one to four years. The goal of young Malay couples is to live on their own as soon as possible, and the delay is due to the necessity to first establish an

TABLE 2.1
Malay Household Membership, 1957 and 1980 (per cent)

	1957	1980
One family nucleus		73.4
Married couple with a parent	75.7	4.6
Parents and children couples		6.2
Other two family nuclei	11.9	4.9
More than two family nuclei		1.5
No family nucleus or one person	12.4	9.4
	100.0	100.0

Sources: Chang et al., 1980: 67; Census 1980: VI. 45-6.

independent livelihood and acquire separate housing. In rural areas of Malaysia and Java, the couple would be given land and agricultural capital by either, or both, sets of parents upon marriage, or helped to clear land or establish a tenancy or share-cropping arrangement. In the urban economy of Singapore, the young are often financially independent before marriage, and the delay in moving out from the parental home is caused by the one- to four-year waiting period for rental or purchase of a government flat. During this period the young couple form a separate family nucleus, and independently earn their food, but are not strictly speaking a separate hearth as they usually give a cash contribution to the parents and share food with the parental unit.

Djamour's research on a fishing village in Singapore in 1949 shows the preponderance of the elementary family as the basis of the household unit. She (1959: 55) found 77 per cent of households to be elementary, while 19 per cent included three generations or other kin, and 4 per cent were composed of single individuals. In a rural Malaysian village, Wilder (1982a: 40) found 68 per cent of households to be elementary in structure (including those comprising married couples without co-resident children), 14 per cent of households were composed of a single adult, and 18 per cent were formed by extensions of the elementary unit to include a third generation or other adult relative. Jay's (1969: 53) findings for Javanese rural households are similar: 74 per cent elementary, 21 per cent extended, and 5 per cent single adults.

While there has been continuity in the predominance of the elementary family in both rural areas of the Malay world and urban Singapore, there have been significant social and demographic changes within the elementary Malay family in Singapore. These changes concern the extent to which the children present in the households are the biological offspring of both the parents. Variations in this are not well reflected in national census data where birth records of children are not checked. The change is due to four factors. First, there has been a rise in life expectancy, making it more common for parents to survive to raise their own children to adulthood. The life expectancy at birth for Malays was 57 years for men and 59 years for women in 1957, and 65 years for men and 66 years for women in 1970 (Chang et al., 1980: 58). Second, there has been a drop in the Malay total fertility rate from 6.3 in 1957 to 1.9 in 1976 (Chang et al., 1980: 59). This again increases the frequency with which parents survive to raise their own children, since they complete their childbearing at a younger age. The tendency to have fewer children has also reduced the frequency with which parents are willing to give children away in adoption, a common practice in Malay societies (Djamour, 1959; McKinley, 1975). The third factor is a reduction in the divorce rate, which in Singapore is calculated as the number of divorces per 100 marriages per year. This rate dropped from 52 per 100 in 1957 to 14 per 100 in 1979 (Wahidah Jalil, 1981: 36). The decreased divorce rate has reduced the frequency of the presence of stepchildren in the home, and it has also reduced the practice of child transfers. Divorce was a prime factor in the decision to transfer a child, since it was thought preferable for the child to be raised by a grandparent, other relative, or even non-relative than to be raised by a stepparent, particularly a stepmother (Djamour, 1959; McKinley, 1975). The fourth factor is the cost of education, which makes the adoption of extra children less attractive. In Singapore, temporary baby-sitting is the most common arrangement in which non-filially related children co-reside in a household. The pre-school child of a working mother is often cared for in the home of kin or non-kin for a fee, and the child is expected to return to its parents on starting school.

Ethnographic accounts give an indication of the magnitude of the change in household composition that has resulted from the trends noted above. Djamour found that in the Singapore Malay fishing village of Tanjong in 1950, 15 out of 48 households contained children who were not the offspring of *either* parent. Life histories collected in Singapore in 1982 revealed that 41 out of 80 individuals aged 30 or above had not been raised by both their own parents up to the age of 15, due to divorce, death, migration, or permanent transfer. This figure excludes temporary transfers of a year or so. The frequency of transfers is much lower among those under 30 years of age.

The impact of these changes on the parent-child relationship and the transfer of resources between the generations will be discussed in a later chapter. Here it has been established that the unit of co-residence in the 1980s is most likely to consist of parents and their own children, though some variations occur. In the following discussion, the term 'household' will be used to refer to the unit of co-residence, and the complexities of provisioning the hearth will be analysed using the concept of householding defined earlier.

Consumption Rights

Co-residence defines a set of consumption rights. These are the rights to share in cooked food, and the use of household items, such as furniture and television set. These are not absolute rights, as consumer goods are considered to be the property of the individual purchaser, placed in the house for the use of others on the basis of a voluntary contribution or gift. There are occasions when a comment might be made linking the consumption of cooked food to contributions towards buying it, but this is rare and it is considered to be very calculating (*berkira*). The conceptual stress is on the sharing of cooked food among everyone present in the house at the time the food is prepared, including informally adopted children and kin or friends who might be there by chance. This is enjoined by Malay hospitality and by Islam.

The boundaries of the regular consumption unit are not clear-cut. Members who do not co-reside, notably married children, often share cooked food. This may happen on a fairly routine basis, for example when a married working daughter, whose child is being cared for by its grandmother, goes daily, alone or with her husband, to collect or visit the child and at the same time eats in the parental home. This daughter's own housing unit is used only for sleeping, and she may cook there only at week-ends. It is often jokingly referred to as a hotel. Having her own housing unit gives the daughter the status of living away from parents, despite the use of the labour resources of her mother. For these services, she would contribute cash (see pp. 44-8). This shows that householding relationships can span across two or more housing units.

Another variation in the consumption unit occurs when a co-residing member makes minimal use of household facilities. This is typically the unmarried son, whose peer activities and voluntary attachments to the households of friends or kin take him out of the house, and whose wage enables him to buy cooked food regularly outside. Alternatively, he may make use of household resources, coming home to eat, sleep and bathe, but rarely be present in the house at other times. Other marginal co-residing members are divorced men without custody of children, who are deprived of the right to separate housing under government housing rules. Though they may eat and sleep in the house of a relative (most often a parent or sibling), their physical presence in the house is likely to be awkward, whatever their financial contribution.

Hildred Geertz (1961: 123) notes that in urban Java, 'single men can attach themselves loosely to a household, take their meals in restaurants, wash some of their clothes themselves: but sooner or later they get married'. This she attributes to their practical dependence on women for the performance of household tasks. In Singapore, too, where men desire the physical and emotional comforts of a 'home', they are forced to enter householding relationships. Commercial substitutes for domestic services are readily available in the city, but are generally thought by Malays to be expensive and inferior in quality. Culturally, the only household relationships that can provide regular current consumption services such as cooked food and laundry are those of spouse, parent, or unmarried child. Other relationships are temporary or uncomfortable, though slightly more tenable for women than for men, as younger divorced or widowed women can sometimes reintegrate with the parental unit.

The social and emotional aspects of home and family are probably even more important than domestic services of a material nature. Few Malays relish the prospect of being alone, even temporarily, and still less as a permanent living arrangement. One divorced man in his early forties who was living alone, experienced a personal, social unease which he described thus: 'Where can I go? I can't stay in the mosque, that will be locked in the evening; I can't associate with the other bachelors who are younger than me and like to go to hotels and bars; I can't visit my friends or sister all the time because they have their own families and want their privacy too.'

Considerable pleasure is taken in family gatherings of the co-resident unit or wider circles of kin (on special occasions). Life histories collected, reveal differences in the tone of such gatherings, depending on the personal characteristics of family members, especially senior males. Some individuals recount extremely strained and largely silent relations with very stern, formal fathers (*garang*), while others claim to have been able to laugh and chat freely with their parents (*gurau*). First-generation migrants from Java are more likely to have been stern, and distant from their children, though the pattern is not entirely consistent. Differences between families still remain, but the trend observed is for parents to foster close and affectionate communication within the family, in part for the sheer pleasure that such close relations bring to all concerned. Strains still seem to arise, especially for boys during the teenage years, but both parents and teenagers state that, ideally, relations should be close and communication open: this ideal is important to the discussions of particular relationships that follow.

The analysis now turns to the question of contributions to the household, to the source of cash and domestic labour that makes food and services available to the consumers. In brief, women do domestic work and all wage-earners contribute some cash, but there is no concept of the household as a unit in which these resources are automatically pooled. It is through an analysis of the dyadic relations between household members that insight is provided into how these various contributions and the relations between them are perceived and organized. The key relationships are those between husband and wife, and parent and child.

3 Householding: Husband and Wife

THIS chapter considers the householding relationship between husband and wife. It discusses the obligation of the husband to maintain his wife, and the wife's obligation to perform household duties, then indicates the ambiguities in the actual exchanges of cash and services which leave each party feeling that they have given cash or services as voluntary acts of generosity. Finally, it discusses divorce and the division of property at divorce, which give further indication of the way the current household exchanges between husband and wife are organized and perceived.

The Obligations of Maintenance and Housework

The clearest principle in the relationship between husband and wife is that of *nafkah* or the obligation of the Muslim husband to provide adequate food, housing, and clothing for his wife. The wife of a non-provider will be advised by kin and friends to seek a divorce, because it is felt that there is no purpose in a woman remaining married if her husband does not maintain her. Failure to provide maintenance is grounds for divorce under Muslim law, and marriage guidance courses and counselling impress upon husbands their obligation to provide support. Under Muslim law as it is enforced in Singapore, the wife is not obliged to participate in the provision of maintenance from her own property or her current earnings. She retains upon marriage all the rights and property she had as a single adult Muslim, and she can sue or be sued in her own name (Administration of Muslim Law Act 1970: 41-3).

The idea that the husband should be the one to provide for the needs of his wife and household can be seen to operate in the household's budget even where husband and wife are both working. The husband's wage is used for essentials such as rent, utilities, education, the monthly order of rice, oil, and milk, and major consumer goods such as the refrigerator, furniture, and television set. The wife's income is used for items that are perceived as supplementary, such as her own and the children's clothing, smaller consumer goods, or goods in her own domain such as a washing-machine and special kitchenware, children's tuition, snacks for the children, and to augment the food budget if it runs short at the end of the month. This allocation of resources is quite meticulous. If, for example, the wife who has cash at hand pays the school fees or the grocery bill, her husband is likely to pay her back. The husband usually sees his wife's wage as her own, and it is often a point of pride expressed by both husband and wife that he never even asks her how much she earns, let alone asks her to give him money: 'Sometimes I want to tell my wife not to spend her money on snacks for the children, to buy better food, but she really loves the children and spoils them. It's her money. She works for it, so I just keep quiet. I have never "eaten" her money.'

Where husbands do ask for regular sums from the wife's wage for their own use or for household consumption, this could cause the wife to seek a divorce. Borrowing or helping out is common in stable marriages when financial circumstances require it. The husband or wife who borrows from the spouse often states the desire to repay the debt, but where the relationship is close, the spouse is not expected to press for repayment.

Household data collected did not suggest that higher levels of income, or better education in either or both partners, engendered a significant change in the view of the husband as the family provider. A budget breakdown provided by a teacher in his late thirties, whose wife is also a teacher earning an identical salary, demonstrates the actual and perceived role of the husband as the primary provider:

I pay for the house, the monthly groceries, the utility bills, school tuition for the children, their religious class, and the car loan and parking costs. My wife pays for fresh food at the market and gives money to her parents [who take care of the four children]. She saves the rest or spends it as she likes. If I run short, I ask her for $200–300, but then I feel like I want to pay it back because I have to ask her so often.

If the wife goes out to work to supplement the domestic budget or to provide 'extras' for her family, this is seen by both husband and wife as her voluntary contribution or gift. It goes beyond her obligation. The wife's clearest obligation is to provide domestic services for her husband and children, and much of her wage goes to pay for baby-sitting, cooked food, and possibly laundry or ironing which substitute for her domestic labour while she is at work.

This is the simple formula for exchange between husband and wife, in which the husband has a clear-cut obligation to provide support, the wife to provide domestic services, and the wife has an unambiguous right to her own cash earnings which she can use for family consumption needs if she wishes, as her gift. There is more ambiguity in the householding relationship than this formula suggests. The ambiguity concerns the management of budgetary margins and the value of domestic labour time, and it is here that the idea of the gift becomes quite fundamental.

Budgetary Margins and the Distribution of Rewards

It was noted above that the first claim on the husband's wage is the maintenance of his wife. He must also maintain his children. Where the wage is low, and the wife is not working, there is a fully egalitarian partnership between them in which he provides cash, she provides domestic labour, and both have an equal right to share whatever food and amenities are available in the home. Where the wife is working, an element of inequality is introduced in that she can potentially save and spend her own money while he may have no margin for his personal use after paying for maintenance. The more common situation is one in which the wife is not working and the husband, as the sole wage-earner, controls the distribution of funds between them. If the husband earns a surplus over and above the costs of maintaining his family, a potential source of imbalance is introduced. The key question then is to determine in whose hands this surplus accumulates and how it is seen.

There are three parts to the domestic budget. The first is the fixed monthly expenditure on rent, utilities, school fees, and debts for consumer goods, and perhaps a regular order of rice, milk, sugar, and oil. The husband may pay these himself, or give his wife the money to do so, whichever is convenient. The second element in the budget is the working husband's monthly allowance for transport, food at work, and perhaps cigarettes. The third element in the budget is for fresh food at the market, a sum usually managed by the wife. The wife is rarely given a separate allowance specifically for her own use, so her access to cash is determined by her management of the variable part of the food budget. She may need to draw on this cash to meet extra expenditures such as cash contributions to weddings, visits to the doctor, and the children's demands for snacks and sweets, and in low-income families an insufficiency of funds is directly reflected in the poorer quality of food towards the end of the month.

Whether it is the husband or the wife who has greater personal spending and saving power depends on how much of the variable budget the husband allocates to himself and how much he gives his wife. The practice varies according to personal and situational factors. One arrangement observed in some lower-income families is for the husband to give his entire wage to his wife and ask back a daily allowance for his food and transport. He may claim to be afraid he might otherwise spend too much and leave the household short while his wife is more skilled at budgeting. He may save out of his daily allowance, or take just what he needs, asking for less if he has a remainder from the previous day. Other arrangements are for the husband to retain a portion of his monthly pay at source and let his wife manage the remainder, or for him to give his wife a monthly sum which he considers sufficient for the market, both patterns being observed over the range of income levels. In all but the first arrangement, both husband and wife have access to some of the surplus, depending on the initial allocations and on their management of their respective budgets.

The monthly allocation of funds between husband and wife is important because once the transfer to the wife has taken place, there is a tendency for that money to be regarded as her personal individual property. The transfer of funds to the wife is seen by the husband and wife primarily as an irreversible gift and not as a joint saving or consumption fund. Both husband and wife like to claim with pride that the husband never asks his wife to return to him, or to account for, any of the money he has given her: 'My husband gives me all his wages, he just keeps his bus fare. If it is not enough, he asks for some back. He calls it "borrowing" and he pays it back. He says, "What I have given you is yours, it is in your hand."'

The money given by the husband is hers to spend or save as she pleases. Both husband and wife consider that when the husband 'borrows' back from the wife, money he had originally given her, he should repay the debt. It is shown in the following example that a wife can also refuse to 'lend' money that her husband has given her:

My husband gives me all his wage ($1,000); I give him $200 a month as his allowance. If he asks me for more, I write it in a book. It is a debt and he has to pay it back. When he went to Bangkok on business I gave him $1,000, and when he returned he gave the balance back to me. A while ago, he had no money and wanted to borrow from me. I said, 'Go and borrow from your mother, she's got lots of money.'

If the husband has given the wife only part of his income and both have separate savings, they negotiate over who will provide cash for any particular purpose, such as the purchase of a large consumer item or the financing of a feast. The decision of the wife to provide cash, whether from her own earnings or from money her husband has given her, is regarded as 'helping' and it is voluntary on her part. Often the transaction is seen formally as a partnership in a particular item, making explicit the fact that two individuals have contributed resources, and the item is jointly owned:

Daud earns $1,200 a month; he gives me $400, $200 I use for food and $200 I save; he spends $300 on the car, $200 on his own expenses, $100 on utilities and the house; he gives $50 to his mother, and $50 to our daughter for her tuition fees, school bus, and her savings account. That leaves him only $100 extra if the car breaks down, so for the car we share (*kongsi*). Like last week, we had a repair bill of $150, I paid $75 and he paid $75.

The practices described suggest that the money given by the husband to the wife, over and above that used to buy food in the market, is regarded by the husband as an outright gift over which he has no further claims. The gift enables the wife in turn to make gifts to the husband and to the household by her voluntary decision to 'help out' with cash.

Although the cash given by the husband to the wife appears to be treated as her personal property in the same way as her personal earnings, it is rendered ambiguous in three ways. First, since the money the husband gives his wife is not separated from market money, there can be a claim by the husband to convert her savings, or the gold and goods she has purchased, back into maintenance money when times are hard. This is not routinely done, but instances were described in Singapore where this was considered legitimate by both parties in the case of dire necessity. Swift (1964: 137) notes that in Malaysia, the money given by the husband is regarded as a family reserve, though he has to convince his wife of the need to use it. Because of this element of ambiguity, the wife often keeps the extent of her savings secret, fearing that the husband might otherwise make claims on that money or else reduce the amount he gives her every month. This would not be the case were her savings based on her personal earnings which are unambiguously hers and should not be converted into maintenance money. In rural societies where gold or land are the principal forms of saving for women, these are not easily kept secret from the husband. However, land and gold are significant assets which the wife would be most reluctant to see converted into current consumption, and the husband would be ashamed to ask his wife to sell them. Cash is kept secretly in Singapore, either in the house or in a Post Office savings account, or sometimes kept with a friend or relative if it is feared the husband may find and use it. Husbands also keep the extent of their savings secret from their wives, whom they fear may make excessive demands.

The second source of ambiguity in the gift of cash from husband to wife is that its origin as a gift makes the wife feel beholden. Though the husband may give the wife more than enough money, so that she has a good margin for saving and personal use, she does not feel she has a full and unambiguous *right* to that money, and feels burdened by a sense of debt and dependence. The wife fears she could be criticized for demanding, and then spending, money that someone else has laboured to earn. As one woman commented, in relation to money given her by her husband or children: 'It doesn't feel good (*tak sedap*) to ask for money, to take other people's money. It is better to have your own, then when you want to buy something you don't have to think twice and no one can say to you, "Oh, that money I gave you, have you finished it already?"'

Thirdly, there is a sense in which the wife considers the money given by her husband not as a gift but as her wage for domestic services. The ambiguity stems from the fact that the husband's wage is regarded as rightfully his, since he earned it by selling his labour power, his own time and energy, to his employer. If the wage is his, then when he gives a part of it to his wife, he is making a gift. Yet the wife also feels she has expended labour and time, and is entitled to some form of compensation, though this is rarely stated explicitly for reasons to be discussed in detail later. Although wives have an underlying sense of entitlement to a share of the surplus from the husband's wage, they like to see themselves as generous in turn, by being moderate in their demands for money: 'I don't like to ask him for money. He wants to give me money but I'm always afraid he won't have enough; he has to pay for so many things. So I rarely ask. I always find out first whether he has enough. I don't like to give him trouble.'

In this statement, the wife is expressing her generosity towards her husband, which is based on her affectionate concern for him and her understanding of his struggle to provide for the family. She also claims that the husband always *wants* to give her money, which is a significant statement about the character of the partners: only a good husband would *want* to give things to his wife, and he would only *want* to give her if she were a good wife. To say he was obliged to give to her would inadequately express the kinship sentiment which motivated the transaction. Goodness, love, and esteem are conveyed by one partner to the other, and to observers, by the assertion that transfers are gifts. These sentiments are not negated or contradicted by the actual separation of the husband's and wife's personal economies, nor by the obligations of religious law or Malay opinion that indicate certain transfers to be right and proper. Nor is the idiom of the gift superficial or false. It is integral to the householding relationship and affects the way that material resources are handled.

The overall effect of these arrangements is that the husband's and wife's resources are not pooled. The resources they each hold separately are not necessarily at the disposal of the other for the financing of business projects or expanded consumption standards. It is not a householding relationship geared to the expanded reproduction of a joint conjugal estate or mutuality. Any mobilization of funds on a joint project needs to be explicitly negotiated, and the proceeds tend to be redivided between the partners when the project is completed.

Domestic Labour: An Undervalued Commodity

The sense in which the husband sees the money he gives to his wife as an act of generosity on his part, was described above. His wife views her domestic services in a similar way. As in the case of her husband's 'gifts', she prefers to see her efforts to prepare his favourite food and to provide a comfortable home as a demonstration of her love and concern for him rather than an obligation. To stress the obligatory aspect devalues the kinship sentiment which motivates her to provide these services in personalized ways and in over-measure.

There is a further sense in which the wife feels she is making a gift of her domestic services. The developed commercial economy of Singapore makes housewives conscious of the cash value of the services they perform without pay, in the home. Many Malay women have been employed in various types of domestic service, and they know that the fee for cooking at a restaurant or stall is $30 per day; washing up is $10; baby-sitting is $80–250 a month, and house cleaning $100–200. While it is the obligation of the wife to perform these tasks in the home in return for her maintenance, their value in commercial terms is more than mere subsistence. If performed outside the home, these services produce a personal income in cash, an income certainly greater than the little 'extra' the wife manages to save from the housekeeping money given by her husband. As one woman bluntly stated, 'Doing housework at home, you don't get paid. It would be better to work outside the home to get money.'

It has been argued that the wife sees her domestic services as a gift, partly as a result of the commercial economy and her awareness of the cash value of her services. She feels that she has given more, and received less, than is really her due. But the claim that these services are a gift is simultaneously undermined by the commodity economy itself. In terms of the commodity economy, domestic services appear to have no product, since they are repetitive and there is no accumulated concrete result. There is no proof of the value of the labour expended, since there is no wage and in a commercial economy, services and products which do not receive a price appear to have no value. The wife who stays at home to cook, clean, and look after children feels, and is made to feel, that she has done nothing. The housewives often express this feeling with the phrase 'I don't work, I only do housework'. This is a feature common to all market economies with a division between domestic and market production and it has been well documented in Women's Studies literature.

The devalued status of housework in the Singapore economic system is the reason why the idea that the husband *owes* the wife a share of the extra from his income as her wage is only expressed jokingly, hesitantly. There is an awareness of the value of domestic services, but this is undermined by the *de facto* status of housework as unwaged. The devalued status of housework is reflected in property division at divorce (see pp. 25-7).

It is not only products and services, but time itself which has a cash value in Singapore. Women are aware of the wages they forgo by remaining at home to care for the family, their awareness further increasing their sense of generosity. Life-history data show that, at least since 1950, many married Malay women have obtained a personal income through their informal sector activities such as selling cooked food, taking in washing, or the practice of Malay crafts and skills. The extent of their labour is underestimated by the official employment statistics for Malay women (see Chapter 7). Women saw their income-generating activities as ways of stretching and augmenting the domestic budget and increasing their personal access to cash. The scope for these informal activities has now declined in Singapore so that women are faced with a more defined choice of whether to work for a formal wage, or have no personal income at all. Though Singapore's multinational electronics assembly companies prefer to employ younger women, even older women find that relatively full employment in the 1980s has made jobs as house-servants and cooks more accessible and lucrative than was the case in the 1960s and 1970s. Wages for the unskilled, lower-income, male population failed to keep up with inflation during the 1970s, and for many families the wife's income is essential to meet basic needs (see Chapter 7). In a majority of families, if the wife did not work, there would be little or no margin from the husband's income for her personal use, a factor increasing the incentive for women to work while also increasing the 'opportunity cost' of those who stay at home.

Few women now positively choose to forgo all formal and informal income opportunities, and the major restraints placed on their labour-force participation are the demands of child care, or the cost of child care relative to low female wages. Muslim husbands have the right to forbid their wives to work, on the grounds that they provide sufficient funds for household needs and the wife is needed to care for children and see to their education, and some husbands fear infidelity if the wife is out of the home; but restrictions are not common, and most wives are free to choose whether or not to work. It is the ability of the wives to exercise this choice, and the known value of their labour time, that makes their unpaid domestic services in the home a form of gift.

It is paradoxical that the commodity economy should reinforce the conception of householding between husband and wife as an exchange of gifts. Domestic services are gifts because they are unpaid though their value is known. They are also gifts because they are personalized and actually unpurchasable, given with loving care and devotion, which means that the money that passes from husband to wife, however generous, cannot be reducible to the wife's pay, and retains the status of a gift in recognition and expression of kinship sentiment. Incommensurability in the exchange means that there are permanently outstanding debts of material transfers and of kindness, and these debts form the basis upon which a relationship is constructed between individuals brought together voluntarily. If everyone were really to calculate, no one would be satisfied, so all parties see themselves as giving generously while still feeling that they are actually giving too much.

Property Division: Separate Economies and the Balance of Power

The division of property on divorce reveals the calculations or negotiations of self-interest between husband and wife, and the relative power of the marriage partners in Singapore's urban wage economy. The economic nexus of the marital relationship can only be directly expressed when the will to sustain kinship sentiment has disappeared and the idiom of gifts has been abandoned, which is the situation of most couples upon divorce. The legal basis for the division of property is rather unclear (see pp. 29-31). In most cases, the property settlement between divorcing couples is made out of court at a meeting between the parties in the office of the Syariah Court clerk. There, each party states his or her claims for property, and negotiations and compromises are made to reach an agreement which avoids the lengthy court process and hastens the registration of divorce.

Since quite often no court proceeding takes place, the rights and wrongs of the divorce itself, such as adultery by either party, are not legally established and so they are not brought to bear on the property settlement. This situation differs from that described by Djamour

(1959: 125) based on her observations in 1949-50, when, in a rather less regulated and bureaucratic process, Malay customs (*adat*) and local knowledge about the circumstance of the divorce were taken into account by the neighbourhood Muslim official in the divorce proceedings. In the ten cases of property division observed at the Syariah Court during the course of this research, the emphasis appeared to be on reaching a speedy settlement rather than on implementing clearly defined principles of distribution. The trends in property division and some of the attitudes expressed are described here.

Household goods tend to be claimed by the purchaser. The wife claims the kitchenware, washing-machine, her clothing, and jewelry bought out of her own earnings, or her savings from market money. The husband generally makes no claims on these, suggesting that gifts from husband to wife are indeed final. Nor is the issue of personal savings in cash or in the bank ever raised, again indicating that gifts are final and that funds accumulated separately are personal property, not a joint fund. There are cases where the husband has left the marital home and already has another liaison outside, so he is not interested in claiming household goods. If the marital home is being dissolved, he claims the major consumer goods such as the television set and refrigerator which he bought. The tendency to divide marital property according to who earned and purchased it caused the court clerk to comment that 'each partner finds his or her own money, and each claims his or her own goods. It is all very calculating.'

The wife does not generally make any claim to the goods the husband bought with his earnings, stating only that 'it is his money, it is his body that works, they are his things not mine'. This attitude on the part of the wife shows that she does not regard her domestic services as an integral part of a joint production unit in which the total joint product, including all his wages, become joint property. The devalued status of housework is expressed also with regard to housing bought out of the husband's wage. Wives comment that they are mere lodgers on their husband's property; they feel they have no right to it as the husband alone earned it while they themselves did not work, but just stayed at home.

It is clear that contributions to householding 'in kind' do not carry the same power as contributions in cash. Government housing rules actually decree that the marital home is the joint property of husband and wife, and that each is entitled to a half share. This fact dismayed some of the divorcing husbands, who asked the court clerk how their wife could possibly have a share if she had not paid a single cent towards the purchase. The regulation requires the party who is to retain the property to renegotiate a loan in order to pay off the other party, and continued possession depends on the ability to pay the monthly instalments. Non-working Malay wives are better protected under this ruling than under Malay custom or Islamic law (see pp. 29-32).

Apart from housing, Central Provident Fund (CPF) savings are the major form of accumulation of Singapore Malays. The CPF is a government-run compulsory savings scheme whereby a fixed percentage of monthly income is deducted from each worker and a similar sum is added by the employer to be placed in a special account. This money can be used only for the purchase of housing or for medical bills and the remainder is realized at the age of 55, intended for old-age support. There is no suggestion in divorce settlements that the wife has any claim to a share of this money earned by the husband. If stably married, Malays consider it a good gesture for a husband to give some of this money to his wife, and often it is used for both husband and wife to perform their *haj* (pilgrimage to Mecca), but this is seen as a gift, not as the wife's right to a share in the household's long-term accumulation. Through the husband's contribution to the CPF, a sum of money that might have been available to the wife as increased market money through the current budget becomes controlled by the husband. Possibly, it will be controlled by another wife if the husband has divorced and remarried by the time the money becomes available at the age of 55.

Where the wife is the party wronged by divorce, the Syariah Court can award a claim for a consolatory gift or *muta'ah*. This is in recognition of her labour services to her husband during the marriage. It is calculated at the rate of $1 per day, which, though low if viewed as a wage, can still amount to considerable sums over a long marriage. The diligence of the court in informing divorcees of their right to claim this money is a significant step in recognizing the unequal implications of waged versus unwaged work.

The description of property division provided here suggests that it was correct not to assume egalitarian, co-operative pooling as the basis of householding. Despite the tendency to describe the relationship between husband and wife in terms of their respective duties, and in terms of generosity and gifts, there is intense calculation of the flow of material wealth between the partners. In the context of Singapore's urban wage economy, the balance of power appears to be in favour of those who earn cash. The individual worker who receives a cash income directly from the employer is able to retain those funds or make gifts as he or she chooses. The debts that women feel are created by contributions of domestic services, are largely unrecognized. This provides a strong incentive for women to work outside the home. The wife's alternative strategy is to make demands on her husband to give generously, allowing her to build up personal reserves of cash as a form of security in case of divorce, and to equalize the balance of financial power between them.

Despite the reduction in the divorce rate in Singapore noted earlier, discussions with Malay women suggest that the possibility of divorce is a factor shaping the householding relationship between husband and wife even where no divorce is currently being contemplated. The principle of separate properties is always present, but the actual willingness of the parties to give generously to each other, rather than demarcate their property, save, and accumulate separately and secretly, tends to reflect the current state of stability of the marriage. The flow of gifts from husband to wife which enables her to accumulate savings and jewelry is a sign of kinship sentiment and of the intention of the husband to persist with the marriage. Djamour (1959: 43), writing of Singapore in 1949-50, notes the tendency for surplus funds to accumulate in the wife's hands and states that the 'average husband values his wife's love too much to deny her the pleasure and pride of wearing valuable jewelry, and he may be unable to resist her persistent nagging; while some men fear that such a denial might precipitate the breakdown of the marriage...'. She also found that husbands related their failure to maintain their children after divorce to the reserves they had given their wives, which should now be used to provide for them. This suggests that, retrospectively, generosity is linked to the possibility of divorce. Swift (1965: 106) also found that for a husband to start accumulating property is a sign that he is planning to end the marriage. In a stable marriage, accumulation is expected to be in the wife's hands, which is both a sign of stability and a reserve for her personally in case of divorce.

Swift's (1963: 280) analysis of the effect of divorce on the householding relationship between husband and wife which discourages the pooling of resources could equally well apply to contemporary Singapore:

The identification by husband and wife of their individual interests with that of the family as a unit is limited.... Each partner, but especially the wife, for it is she who has the greater fear of being left unsupported, is concerned with the improvement of the economic position not only of the family as a unit, but also of his or her own individual position within it.... If a wife helps her husband to get rich, how can she be sure that he will not use this money to attract another younger and prettier wife, or that they will not quarrel and so herself be divorced?

The wife invests in land in her own or her children's names, and discourages the husband from risking family capital in business. Swift (1963: 280) continues: 'Wives may devote all their resources, even raising more from their kin, and bear great discomfort for the furtherance of their husbands' ends, but as marriage is defined in Malay society they are not required, or even expected to do so.'

A life history collected in Singapore illustrates several of the themes discussed above: the perceived precariousness of the most stable marriage (subject to even magical destruction); the non-providing husband who is a drain on his wife's resources; secret savings, and the preference for non-liquid assets which are resistant to being spent or borrowed by others, including the husband.

Mina came from a poor Javanese family in Batu Pahat, Malaysia. On marriage, she and her husband both worked hard and accumulated 50 acres of land. A significant source of income which went into the joint accumulation of property, was her earnings as a midwife, traditional curer, and bride attendant. The marriage was happy and they had seven children. Then a neighbour became intensely jealous of the couple's material success and of the happy marriage, and put a spell on the husband so that he would forget his wife and think only of the other woman. He changed character and started to complain about everything, including Mina's cooking which hitherto he had found excellent. He pronounced divorce and married the other woman, who brought along six children of her own, so that he was 'feeding other people's children'. An agreement was reached to divide the joint household property among the children, Mina retaining a 5-acre plot. Mina moved to Singapore to escape village gossip and to avoid further contact with her ex-husband and his new wife. For a long time she resisted offers for remarriage, stating, 'You start as one, go back [die] as one, better to live as one.' She finally remarried. Her present husband earns a very low wage as a security guard, and she earns money by selling cooked food. The husband gives her $100 a month, which is insufficient to cover rent ($80) and utilities ($80) besides food, but he never asks her whether she has enough money, assuming she will make up the difference. In addition, he takes $4 per day from her for his cigarettes and personal use. She is trying to save secretly for her own old age (not explicitly against the possibility of divorce as he is not significantly supporting her anyway). Her husband discovered her savings book and forced her to withdraw all her money to lend to his relative. She has started saving again, keeping the receipts more carefully hidden. Her husband has been trying to persuade her to sell her land in Batu Pahat. She refuses to change this asset into liquid form because, especially in her husband's hands, 'money goes quickly' and he is not giving her any cause to believe he has considered her old age. She tells him, 'This is my own property from my previous marriage, not property acquired together with you, you have no right to it.' Her strong and 'ungenerous' stance may jeopardize the marriage further, but she feels she must protect her own interests.

Householding: The Malay Cultural Heritage and Islamic Law

Since the consideration of property settlement at divorce proved to be a useful source of information on the householding relationship between husband and wife, and since the possibility of divorce itself seems to be integral to householding, it is this theme that will be pursued in looking at other parts of the Malay and Javanese world. Comparison of other Malay areas with the situation in contemporary Singapore helps to identify the common cultural elements as well as indicate the significance of Singapore's economic conditions in shaping the marital relationship.

In the late nineteenth century, Maxwell (1884: 125) noted in Malaysia the principle of *harta syarikat* under which husband and wife are entitled to equal shares of the product of their joint labours. *Syarikat* is an Arabic word meaning 'the joint earnings of husband and wife; a partnership; a mercantile firm' (Coope, 1976). An alternative expression is *harta sepencarian* which could be translated as 'wealth found by people working together'. The contractual partnership of husband and wife in seeking specific items of property has been the conceptual basis of property settlements on divorce, or the death of a spouse, throughout the Malay world. The interpretation and implications of the contractual partnership principle have differed depending on the economic circumstances.

In a rice-growing economy, husband and wife labour jointly on one product. The customary division of labour gives both the husband and wife essential roles to play at different stages in the rice cycle (Wilder, 1982a: 28). Their partnership in production is reflected in the division of marital property in most rice-growing areas, which generally follows the principle that property acquired during a marriage by joint efforts in rice production should be equally divided. The property of husband and wife before marriage remains separate, and on death reverts to the children or parents of the individual, not the spouse (Jay, 1969: 63; Swift, 1963: 275).

As soon as husband and wife labour separately on different products, it becomes clear that the basis of *harta syarikat* is a calculated partnership in a specific product or venture, rather than an indivisible joining of their labours in general. In his discussion of rural North Peninsular Malaysia, Banks (1976: 581) defines the basis of division of *harta syarikat* as involving a precise breakdown and calculation of the capital and labour that each partner had contributed to an object such as an addition to a house: 'One may determine the size of individual shares in jointly acquired property through an investigation of initial and cumulative investments in labour.' The idea that not only the initial capital but also the labour power and labour product of husband and wife remain their separate property during the marriage (though sometimes combined in a specific joint venture such as a major purchase or jointly raised crop), opens the way for the devalued status of housework identified in contemporary Singapore. If the underlying principle is 'to each the product of his own labour', then the property relationship between husband and wife in a rice economy is egalitarian only because of the culturally established division of labour and production requirements pertaining to that particular crop, and not because of any conception of the husband and wife themselves as an integrated unit of production and ownership. In an urban wage economy, the extent of the home-maker wife's contribution to the production of her husband's wage is unclear and there are various interpretations.

In a legal case in Perak in the 1900s, the court found that the wife had *no* claim at all on her husband's money earnings while she was at home (Ahmad Ibrahim, 1978: 271). In Singapore in 1950, Djamour (1959: 39) found that the house and furnishings bought with the husband's earnings belonged jointly to the husband and wife, while the wife's personal earnings and gifts from her husband, were hers alone. She does not state the ownership of that part of the husband's earnings which was not realized in the house and furnishings. Elsewhere (1966: 4), she writes that the principle of division of *harta syarikat* is one-third to the wife and two-thirds to the husband where the property was acquired by the husband's sole efforts while the wife was at home.

A legal statement on the principle of *harta syarikat* was made in Singapore in 1978 as the result of an appeal against a ruling by the Syariah Court (*Syariah Court Appeals* Nos. 1 & 2, 1978). The ruling stated that the principle of *harta sepencarian* had been applicable in Singapore since the enforcement of the Administration of Muslim Law Act in 1968, while prior to that date the court had upheld English law in finding that the joint earnings of husband and wife were the property of the husband. The principles are that 'for a property to become *harta sepencarian* it must be jointly acquired by the spouses or by their joint efforts'. Property acquired by the sole efforts or resources of the husband, or property given as a gift to the wife, is excluded from this category. The general trend of division for property established to be *harta sepencarian* is a half-share, depending on the particular circumstances. It is clear from the details of the ruling made, that the principle of *harta sepencarian* refers to the division of a particular piece of property into which both had put resources (for example, $10,000 each towards a house). There is no suggestion that half of all the husband's or wife's earnings during the marriage should belong to the spouse on the basis that they both had contributed to the joint endeavour of householding. This confirms the argument made earlier that the running of the household is seen as a day-to-day, non-cumulative venture, and the bulk of the husband's and wife's resources are kept separate from it. Such transfers as are not consumed in the household become gifts between husband and wife, absorbed into their separate estates.

The principles of distribution under Muslim law are not greatly different from those of *harta syarikat*. The court ruling referred to above used various Islamic texts to derive the ruling that 'in the event of a dispute between husband and wife (on divorce) with regard to properties acquired by them after their marriage (including their matrimonial home) and if neither of them could clearly prove his or her ownership, then both parties are to take oath and the property in dispute is to be divided equally between them' (*Syariah Court Appeals* Nos. 1 & 2, 1978).

Jay (1969: 64) found that in the predominantly rice-producing economy of rural Java, though most people felt half-shares were fair, some who were more inclined towards a canonical version of Islam (the *santri*) favoured two-thirds to the husband in view of his greater responsibility to provide for his wife's maintenance. The majority view that favoured half-shares reflects the fact that in the Javanese rice economy, maintenance is actually provided by the joint efforts of husband and wife, not by the husband alone. The idea that the husband is the provider of food and housing for the wife reflects the urban economic origins of Islam and the assumption of a division of spheres of economic activity between husband and wife. Review of the principles and practices of property division at various times and places in the Malay world, permits the observation with which this chapter began–the prominence of the concept of *nafkah* in the contemporary householding relationship between husband and wife in Singapore–to be understood in proper cultural and economic perspective. In Singapore, greater exposure to canonical versions of Islam among literate city-dwellers, together with an urban economic structure which pays the husband a wage while defining women's work, whether inside or outside the house, as subsidiary, promote the currency of the idea that the *husband* should provide for the household. Simultaneously, the wife's perceived contribution to householding, and hence her property rights upon divorce, are undermined.

An interpretation of householding in an urban environment which is at variance with that offered here is provided by Hildred Geertz. Discussing the poorer sectors of a town in Java, she (1961: 49, 50) stresses the indivisibility of the husband-wife unit despite their involvement in diverse spheres of economic activity. 'Since husband and wife are an economic unity, even though the wife may not participate directly in the acquisition of income, her performance of household tasks is considered part of the productive enterprise.... [Neither party] has any claim to any piece of property gained during the marriage period by virtue of his working alone for it.'

In her discussion of economic relations between husband and wife, she (1961: 125) asserts that where they earn income separately, they do pool it: 'All that either couple earns during the marriage... is the joint property of both of them, to be used for the welfare of both partners and their children. This equality of partnership is understood from the beginning of the marriage.'

This statement contrasts with the findings presented here for contemporary Singapore, and with Djamour's (1959: 39) discussion of Singapore Malays in the 1940s. Unfortunately, Hildred Geertz does not provide data on the division of property on divorce (whether amicably settled or contested), or the treatment of 'secret' separate cash savings, as this could provide a valuable source of additional information. Although conjugal co-operation may be verbally stressed in a casual way–'yes, we work together'–and though some pooling of resources does take place, the evidence presented here for Singapore and other comparative data examined, indicate an underlying separation of husband's and wife's economies. That this separation was also manifested in urban Java in the 1950s, is corroborated by other evidence given by Dewey and by Clifford Geertz on the same town in which Hildred Geertz worked. Dewey (1962: 33) notes that, in view of the high divorce rate, spouses avoid business involvements with each other or with in-laws. Clifford Geertz found that husband and wife operate separately in the bazaar and 'regard each other in that context as coldly as any other trader'. It seems unlikely that if husband and wife will not do business together, they are operating a fully joint economy either in principle or in practice.

No evidence was found in Singapore to suggest that husbands and wives of Javanese origin are any more 'joint' in their householding relationships than Malays. Jay's (1969) ethnography of rural Java also gives no such indication. The practice of householding in Java in the 1950s certainly should not be expected to be identical to that in Singapore, given the different cultural and economic conditions. It is doubtful, though, that the element of negotiated self-interest is completely absent in the Javanese relationship. If it is absent, it would imply that husband and wife have an economic unity wholly defined by their marital status, rather than a temporary and contingent unity achieved by individuals who have agreed to co-operate but who do not expect their futures to be necessarily linked.

The difference between Hildred Geertz's findings and those presented here for Singapore are instructive, and confirm the usefulness of cross time-space comparisons as well as the importance of attention to specific historical and economic circumstances. Among the petty urban traders

Hildred Geertz studied, it is very likely that income opportunities for women are similar to those of men, and this making the concept of *nafkah*, or the husband's obligation to support the family, less relevant to them. The weakness of the *nafkah* concept is concurrent with the low level of literacy and Islamic knowledge, or actual hostility towards Islam, among the poorer urban Javanese (known as *abangan*). In the particular economic and cultural context described by Hildred Geertz, it is likely that husband and wife feel an equal obligation to put resources into the current consumption of the household, as they would have done as equal contributors to the stock of rice in the rural Javanese economy. This should not be misinterpreted as a total pooling of their personal resources. In the rice economy, property acquired during the marriage is equally shared only because husband and wife have been jointly engaged in producing one integral crop. In the diversified urban economy of Java, as in the case of contemporary Singapore, one might suspect that much of the surplus beyond the current consumption needs of the couple and their children, is not only earned separately but accumulated separately, never to be pooled or later redivided on divorce.

Divorce

The foregoing discussion of the householding relationship between husband and wife in contemporary Singapore showed it to be made up of duties and rights, a negotiated partnership based on self-interest, and kinship sentiment, which is the willingness to forgo direct self-interest and forge a relationship out of material exchanges seen as gifts. The Malay language captures the idea of the mutual adjustment or 'fit' of two individual souls in the concept of *jodoh* which is an important element in a successful marriage. *Jodoh* is seen as mystically bestowed, though only retrospectively. Mutual adjustment is known to be a continuous process throughout a marriage, though when a marriage fails, it shows that *jodoh* has now gone or perhaps was never really present. *Jodoh* also applies to the relationship between parent and child, and a child who is often ill or unhappy may be given away in adoption on the grounds that it was not *jodoh* to the parent.

The idea of *jodoh* is related to the concept of the individual as a being upon whom God has bestowed a unique character (Banks, 1983: 68). Parents eagerly look for signs of individuality and special traits of character in children, and are tolerant or even encouraging of their individual likes and dislikes, such as particular foods that they claim make them ill (Banks, 1983: 84). This stress on individuality is the counterpart of the economic individualism discussed extensively above. People are not expected to be able to get along or work together by virtue of their status (husband, wife, parent, child, elder or younger sibling). Their individual characters may make them unable to adjust to each other. Where a successful adjustment does occur, it is both a human accomplishment achieved through their willingness to give and to tolerate, and also a mystical conjuncture. The recognition that two individuals may not be able to adjust to each other is the reason why divorce in Malay society is never really unexpected.

Wilder (1982a: 70) adopts a useful distinction between the normative instability and the statistical instability of Malay marriages. His findings in rural Malaysia, the observations of the contemporary householding relationship in Singapore recorded here, and numerous other ethnographic accounts confirm that normative instability, or the recognition of a potential for divorce, is a constant feature of Malay marriage. As a measure of statistical stability, the ratio most commonly used is that of the number of divorces per 100 marriages per year. By this measure, Java at the time of Hildred Geertz's study and, more recently, Singapore up to the 1950s, and much of the Malay world, have experienced a divorce rate of about 50 per cent, or 50 divorces for every 100 marriages per year (Hildred Geertz, 1961: 69; White and Hastuti, 1980: 16; Gordon, 1965: 26). This is an unsatisfactory form of divorce statistic, as it is affected by the size of the age cohort and by the changing age at marriage. It does not show the proportion of the population affected by divorce, and the high figure could reflect the repeated divorces and remarriages of a small percentage of the population, while the majority of marriages remain stable (Swift, 1965: 120, 131). It could also reflect the instability of marriages in the first two or three years, while after one or two 'trial' marriages or false starts, most individuals then settle into stable unions. In order to overcome the limitations of this statistic, it would be necessary to obtain longitudinal data on the marriage profiles of the whole population, rather than just those registering divorce within a particular time frame. This data is not available for Singapore, but data collected by Wilder on a village in rural Malaysia in the 1960s provides some useful insights which may be of relevance. Wilder (1982a: 71-2) found in his sample of 341 village marriages, both extant or terminated, that 162 or 48 per cent had ended in divorce, and 69 per cent or 98 of the 162 divorces had occurred after less than three years of marriage. This data led him to conclude (1982a: 73) that 'there are two categories of Malay marriage–a brief, tentative type of union, and a longer effective one'. When the divorces occurring within the first three years of marriage are excluded from the sample and the pattern over the remaining years is compared with that of other cultures, the Malay divorce rate is not exceptionally high, so that there is a 'normative' but not 'statistical' instability. Wilder (1982a: 73) also recorded the proportion of the population affected by divorce in his village and found that 40 per cent of men and 42 per cent of women had been divorced at least once.

For Singapore, only the crude statistic of the number of divorces per 100 marriages per year is available, and this shows a decline in the rate from 52 per 100 in 1957 to 14 per 100 in 1979 (Wahidah Jalil, 1981: 36). Some detailed figures obtained by Wahidah Jalil (1981: 63) from examination of the records of the Singapore Muslim Matrimonial Court show that 40 per cent of divorces registered in the court in the period 1969-80 took place before five years of marriage, but this does not show the percentage of *all* Singapore Malay marriages that ended before five years.

The rapid decline in Singapore's Malay divorce rate began with the introduction of the Muslim Law Act (1957) which enforced stricter controls on the registration of divorce. This Act placed some bureaucratic hurdles and financial costs in the way of quick and thoughtless divorce and it seems to have had an immediate effect. The Act also controlled polygamy and age at marriage, both of which may have had a stabilizing effect. Government housing regulations under which single people are ineligible for accommodation add a further disincentive to divorce, as the partner without custody of children is often left homeless. The economic vulnerability of non-earning Malay wives in the Singapore urban wage economy, which was discussed earlier, may constrain some women to remain in marriages which their financially independent, property-owning rural cousins would have terminated.

Observations at the Syariah Court confirmed that legal regulations, accommodation, and economic livelihood are among the pragmatic factors considered by individuals when considering divorce. Besides practical considerations, cultural changes in Singapore have affected the rate of divorce and the way that divorce itself is perceived. The ideal marriage is harmonious and stable, and such a marriage is a source of envy to others (so much so that some consider the magical deeds of jealous onlookers to be the cause of their own or other people's divorces). The distinction drawn by Wilder between early, unstable unions and subsequent, durable ones, though perhaps statistically valid, is not reflected in the views expressed about marriage and divorce in Singapore. Where Wilder (1982a: 75) found that no stigma attaches to the break-up of a marriage, even a first one, the first marriage in Singapore is always entered with the full expectations that it will be the only one. Young people observe divorces among their peers but still tend to expect their own experiences to be different.

A major reason for the expectation that a marriage made in the 1980s will endure, is the fact that few marriages are still arranged by parents, so that the young people feel both optimistic of and responsible for the choices they have made. Indeed, one reason parents give for their unwillingness to arrange their children's marriages is their fear of being blamed if the marriage fails. Depending on the relationship between parent and child, some parents play no role in selecting their child's spouse, while other parents are important as advisers and through moral pressure can effectively prevent a marriage of which they disapprove. No cases were observed or recounted of young people in the 1980s being forced into a marriage for which they felt unprepared or where the future spouse was unknown, disliked, or not actively chosen and approved by one means or another. Such instances were more commonly recounted by the generation aged over 40 (i.e. born in the 1940s or before). Parental concern to marry daughters while young and virgin, and the over-eagerness of some parents to stage large marriage feasts for prestige reasons, led in the past to the selection of unsuitable marriage partners (Hildred Geertz, 1961: 70; Djamour, 1959: 130), but this is no longer the case in Singapore.

A second significant change in the cultural conditions for marriage and divorce in Singapore, not unrelated to the first, is the emphasis placed by Muslim educational organizations, and the Syariah Court itself, on the responsibilities of both partners in securing a harmonious marriage. Marriage preparation courses are run by a number of Muslim organizations and mosque committees, and they produce pamphlets listing the responsibilities and desirable attributes of each partner. These courses are more effective in reaching the better-educated youth with at least secondary education though there are plans to extend them more widely or even to make them compulsory. During marriage counselling at the Syariah Court, it is questions of rights, duties, and responsibilities that are most often stressed both by the counsellor and the partners. These include not only financial responsibilities but also social ones, such as the requirement to be present in the home a reasonable proportion of the time, to talk to each other and take an interest in each other's problems, and to respect and like each other's families. These issues, rather than vague dissatisfactions or questions of mystical harmony (*jodoh*), are those that are seen in the 1980s as critical to a successful marriage, and observers attach blame to one party or the other according to these criteria.

There are obvious problems in stating the cause of divorce since there are always a number of factors involved. Observation of the counselling of 36 cases of marital breakdown at the Syariah Court provided some insights into the immediate precipitating factors and deeper causes of divorce, and below is a summary of the types of cases observed.

The main marital problems in the early years of marriage concern relations with kin and domestic finances. Some of the divorces in later life are linked to the family development cycle as wives who have been unhappy, badly treated, or given insufficient funds for many years provoke their husbands into divorce once their children can work to support their mother. Mental or physical illness is the cause in a few cases. Interestingly, infertility does not figure as a major ground for divorce. This is because the primary reason for marriage in bilateral, individualistic societies is not procreation, but the marriage bond or partnership itself (Macfarlane, 1986). Though infertility is a valid reason for divorce under Muslim law, other ethnographers have noted that in Malay society, an infertile couple is expected to adopt, rather than to divorce for the reason of infertility alone (Djamour, 1959: 93; Banks, 1983: 99). Infidelity, excessive jealousy, and physical violence occur as part of the complaint in the majority of cases, and about 10 of the 36 cases observed seemed to have these as the primary reasons for breakdown.

Problems with kin, leading to divorce, are centred on the practical and emotional obstacles to forming an elementary family unit. Fourteen of the 36 cases of divorce observed, focused on this issue. These are cases where the husband or wife fails to focus attention on their nuclear unit and continues to give time, attention, and resources to their own parents or to children or grandchildren from previous marriages. Banks (1983: 102) found that in rural North Peninsular Malaysia, potential conflict with in-laws is regarded as inevitable, because 'either spouse can suspect that prior bonds of coresidence will take precedence over marital duty'. For this reason, a separate residence for the new couple is considered essential for a stable marriage (Banks, 1983: 141). The delay of 1–4 years before a young Singapore couple can rent or purchase government housing for themselves forces them to stay with in-laws, often in very crowded conditions. Private, rented housing is too expensive for average and low-income earners. The housing situation places a strain on the relationship, as the spouse who feels uncomfortable in the in-laws' home tends to return frequently to his or her own parents' home. This can lead to the couple living separately, and eventually to divorce. Few Malays are willing to contemplate permanent co-residence with in-laws.

In one case observed, the couple were staying with the wife's parents in a one-bedroom flat, with her unmarried sibling and a widowed sister with five children. The marital complaint was that 'the husband always went back to his mother's home'. Poverty plays a role since private, rented accommodation would be available to those with higher incomes in order to solve such an obvious cause of marital difficulty. In two cases, a son had bought a flat for his widowed or divorced mother and siblings, and he was thus unable under government housing rules to apply for a separate flat for himself and his wife. Several far-sighted parents had carefully avoided such problems by securing separate accommodation in the expectation that permanent residence with an in-law would *not* work out, whatever the initial good intentions.

Ties with kin outside the elementary family can still disrupt the marital bond even where the couple has been able to establish a separate residence. In two cases observed, parents who had disapproved of a marriage were accused of using magical means to cause a divorce. The magic was thought to be directed at winning back the emotional loyalty of the child to its parent, particularly the mother. In three cases of second or third marriages in old age, the husband complained that the wife was 'too busy with her grandchildren' at the expense of attending to him and to his own children. Jealousies over attachments to kin outside the current nuclear unit are nearly always phrased in terms of both kinship sentiment and material flow of goods. For example, a wife complained that her husband did not love her but only cared about his mother, because he always went to visit her and always gave her money. It seems that the suspicion of divided emotional loyalties is more disruptive to the marital bond than the diversion of cash and gifts away from the wife, but since the willingness to give material goods is taken as evidence of an emotional bond, the two aspects of the relationship are closely linked.

Many divorces are caused by problems over domestic finances. A newspaper report on Muslim divorce found that men complained that their young wives were selfish in expecting to keep their wages to themselves while the husband struggled to provide maintenance (*ST*, 6.8.84). The more common problem in the cases observed was the husband's failure to provide maintenance. This accounted for 12 of the 36 cases observed. Eight cases involved young men with very little education who were in jobs with low or irregular income. Four of these young men had not themselves provided the customary sum of money (*belanja kahwin*) given to the bride on the day of the marriage, but had borrowed it from their own or, in some cases, their wife's parents, thus beginning the relationship with in-laws badly. The average rate for *belanja kahwin* in weddings observed in 1983 was $1,000–3,000. It happens frequently that a price has been agreed and marriage arrangements finalized, then at the last minute the young man is unable to produce the promised cash, forcing others to cover up to avoid public embarrassment. The young husbands appear to be continuing a bachelor life-style, oriented towards their peers and activities outside the home rather than to the new nuclear unit.

It is a common pattern for young Malay men to take a few years to settle down to the responsibilities of marriage, and they often retrospectively declare that they did not fully realize their responsibilities until the birth of their first or second child. They describe a transformation in their social and working life at this point, whereafter they spent more time at home, broke off ties with peers, and made concerted efforts to hold a steady job. Some young men begin to focus their attention on their new nuclear units sooner than others, and the early years of adjustment do not always lead to divorce. The wives declare that they are patient and try to point out their husbands' responsibilities, though too much of this counts as nagging, or 'always asking for money', and can be counter-productive, driving the husband back to his peers for escape. A married man described his own transformation from bachelor to adult and portrayed the ideal behaviour of a married man thus:

Since I married sixteen years ago I have never been out, never come back from work and gone out alone. If I go out at all, it is with the family or to religious class, that's all. If not, I stay home. I realize that I have a household; this is my responsibility. If I go on like I did as a bachelor, drinking, dancing, smoking *ganja* [marijuana], my household could not be peaceful. I know others are different, those who think of their own happiness. They can be happy but their wife and children are in trouble.

Four of the divorce cases observed, involved older men who since marriage had never supported their families adequately. This was always connected with low education and irregular work. The wives had struggled to support the family, and in some cases had been obliged to give the husband spending money in addition. The husbands deserted the home, either returning only to sleep or actually sleeping elsewhere. They attributed their absence to incessant nagging, and appeared to be seeking escape from responsibilities with friends, drink, or drugs. When at home, they were often violent, as if attempting to reassert authority over the wife and children who were nominally under their care and control. These men did not measure up to their own or their families' idea of the responsible, providing husband and father. In one case, the wife, a seamstress, had supported the children and had been forced to give the husband daily pocket money. In another case, the husband had never paid much attention to his family, preferring to eat at his mother's house next door. In a further case, the wife worked as a house-servant while the husband worked irregularly, gambled, took drugs, and liked to relax and play the guitar with his friends. He was always out of the house. The wife said, 'I have to do everything; I applied for this house, I pay for everything, he takes no interest in the children. What's the use of being married? I'm better off without a husband.' In another case, the husband had frequently deserted the home for years on end. The wife felt that whenever she had 'raised herself up', taken her gold out of pawn, bought herself some nice clothes and was no longer ashamed in front of her relatives, he would come back and again become a drain on her resources.

The pattern of an irresponsible or non-providing husband in a situation of poverty is by no means unique to Malays. There are numerous Singapore Chinese families with similar problems, though the husbands may adopt a different set of vices (alcohol, opium, gambling), and are more likely to desert the home and live with a mistress or second 'wife' than actually divorce (Leong Choon Cheong, 1978; Salaff, 1981). In the Malay case, divorce is the more common option, since it is regarded as culturally and religiously valid in these circumstances. Hildred Geertz (1961: 46) argues that the *de facto* absence of the husband in some Javanese families makes the family system matrifocal:

... the dominant status of the women in the family, the attitudes of restraint and avoidance between the men of the family, the contrasting closeness of the women of the family, the mistrust of affinal relatives, and the strong economic position of women–all work together to produce a matrifocal pattern of familial relationships which is most clearly evident in household composition, one of a solidary core of related women with a loose periphery of the men of the family.

The observations made in Singapore suggest that the statistical presence of households with absent or marginal males does not necessarily mean that this is a culturally validated ideal (see Liebow, 1967). The idea of the husband as the provider of family maintenance and the idea of the married pair as the core of the household are conceptually central to the Malay householding relationship in contemporary Singapore, and other arrangements are always seen as temporary and less satisfactory. The husband's failure to provide finance and some active leadership in the household is not acceptable to Singapore Malay wives. Though the wife may succeed in providing for the household, her own opinion and that of Malay society and Muslim law is that while married, she should not be obliged to do so. It is thought better to be unmarried or to remarry than to remain in a nominal marriage in which the flow of kinship sentiment and material goods, by which the householding relationship is defined, are absent. A marriage becomes stable when these exchanges are established.

The shift away from emotional and financial involvements with parents towards involvement with spouse and children, is essential for the stability of marriage as it is defined in Malay society, but it is still to parents that husbands and wives turn for housing, care of children, and financial aid upon divorce, so the parental bond cannot be neglected. The procedures for enforcing the payment of children's maintenance by the father are available in Singapore, and more women are taking advantage of them. In awarding maintenance, the court has to take into account the husband's ability to pay, especially if he has remarried and has other children to support, so maintenance is often inadequate. There is no developed state welfare system that provides for mother and children in such circumstances, though very small sums may be provided by the government or by charities. Since most Malay couples have a baby in the first year of marriage, nearly all divorces in the first years of marriage involve young children, and divorce continues to be a disruptive factor in the upbringing and education of children.

It is to the relationship between parents and children that the discussion now turns. Husband and wife have a joint responsibility for the upbringing of their children, this being perhaps their most corporate endeavour. Here it has been argued that the exchanges between husband and wife are oriented to meeting daily consumption needs or current reproduction. The flow of 'extra' cash and services builds up kinship sentiment which is a sign, though not a guarantee, that the relationship will persist. In examining the householding relationship between parents and children, the focus is on the longer-term exchanges which achieve extended reproduction across the generations.

4 Householding: Parents and Children

THE issue examined here is the terms on which goods and services pass between parents and children in the current household and over the longer family life cycle. The chapter explores how the changed economic circumstances of old and young, male and female, have penetrated family relationships and contributed to the development of new social practices which differ substantially from those of the rural Malay household, while retaining important links to the broader cultural heritage.

Attention has been drawn to the flow of wealth between parents and children by Caldwell (1976), who is interested in the demographic implications. Caldwell argues that the primary cause of high fertility in some family systems is the value of children as the creators of wealth for their parents. Macfarlane has related the long, historical trend of low fertility in England to children's rights over their individual property and to the absence of a corporate economy between parents and children which ensure that children are generally the creators of wealth for themselves, not for their parents. In the English family system, children are a luxury, or if too numerous, a burden that can reduce the parents to poverty, not an economic asset (Macfarlane, 1978 and 1986). Here it will be shown that the Singapore Malay family pattern lies between these two extremes. Parents can derive some return from children, but the return is conditional and negotiated. Children are a possible asset, but not really a sound investment. Any financial return parents derive from their children is dictated not by a joint economy between parent and child, nor by any defined obligations of children towards parents, but by the emotional bonds of kinship sentiment that have been built up between individuals through a currency of gifts that incur indefinite, but nevertheless powerful, obligations.

Several writers have noted that the flow of wealth in Malay society is from parent to child, and there is a marked absence of emphasis on financial return from children. Djamour (1959: 166) puts this point clearly:

> Parents must be forever ready and willing to help their children in every way (even after they have married and settled in independent residences), and they expect little in return. In principle and in practice there is only slight respect for

old age, and a man is always expected to consider his own wife and children's welfare before that of his parents. An elderly couple, unless they are infirm, must attempt to earn their living, and even if in the past they have shown great generosity toward their sons and daughters, they do not believe themselves entitled to regular financial help in return. In other words, there is hardly any personal reciprocity between parents and children: for the former it is always more blessed to give than to receive.

Djamour's elderly informants in Singapore in 1950 did not feel this was a new state of affairs, or a breakdown in relations between the generations caused by social and economic changes (Djamour, 1959: 144). Islam also stresses the responsibilities of parents towards children and has no concept of nurturing children with a view to obtaining a long-term financial return on investment: the reward for carrying out one's duties is in heaven.

Rural ethnographers have noted an emphasis on the flow of nurture and material goods from parent to child, which is seen both as obligatory, dictated by Islam and moral law, and as a free gift, voluntarily bestowed with no thought of return. Religious authority is felt to be necessary in order to enforce the obligation of parents to care for their offspring, since not everyone is expected to be naturally a responsible parent (Banks, 1983: 64). The voluntary aspect of the parent–child relationship is fundamental, since any adult who is willing to take on a nurturing role towards a child and to help bestow upon that child the means of leading an independent, adult existence is a socially recognized parent (Banks, 1983: 137). Malays often refer to their 'mothers' or 'fathers' and enumerate all those who had a hand in raising them to adulthood. A person's commitment to nurture and care for children is definitional of parenthood in Malay society, and this makes the relationship between all parents and children a conditional one that needs to be individually created and sustained (McKinley, 1975: 62). While consanguinity exists *de facto*, parenthood does not. It was mentioned earlier that inheritance, or the passage of property between the generations, is seen as a voluntary transaction or gift. Banks (1976: 576) writes that '*harta pesaka*, the Malay term for inherited wealth, is the material manifestation of the flow of nurturant, affective sentiments from parent to child'.

The following analysis traces the flow of goods and services between parents and children at the different stages of the family life cycle. It shows that the idea of the gift is prominent in daily life, but this is not to be interpreted as a simple continuation of the cultural tradition scribed above. Detailed examination of the Singapore situation and comparison with other times and places in the Malay world show that there have been deep changes in the relationship between parents and children. Behind apparent cultural continuity, there is the creative production of new forms of life.

The Care of a Young Child

Singapore Malay adults find great pleasure in the presence of small children. This is a cultural feature that has been noted by many writers on South-East Asian societies, and it goes some way towards explaining why children have been desired in abundance although they are not seen primarily as creators of wealth for their parents. Parents whose children have grown up often adopt an additional one just because there is no *small* child in the house, and the home has become uncomfortably quiet. Friends or neighbours borrow small children for a few hours or a few days simply for fun, and they regard the expenses they incur, buying the child snacks or even clothing, as pure gifts. The daily physical care of young children was a less significant drain on the labour resources of rural mothers who could entrust them to older siblings or neighbours' children. This has changed in Singapore now that older children are in school and the mother's time is a more precious commodity. The increased monetary value of mothers' labour is one factor in the drastically reduced fertility trend in the last twenty years. Conceptually, still, one purpose of having children is the pure pleasure that children bring to adults, and nurture is seen primarily as a pure gift.

There is an idea of long-term return for the nurture bestowed on a child, though it is not made central to the relationship. It is expressed in the form of a vague, conditional, distant future in which the child who has been nurtured when young, may in turn care for the parent when old. The terms used to refer to the care of the old are the same as those used for the care of the young (*bela, jaga, pelihara*), and the image is that of a reversal of the parent–child roles. Sahlins, quoted earlier, uses the image of the mother suckling a child to portray his notion of 'generalized reciprocity', where the gift of milk incurs a debt but the expected return is not specified or stressed. Though he appears to regard it as universal, Sahlins' example seems in fact to be derived from the Western family system, since there are societies with joint family economies in which parents are quite explicit about the direct economic returns they expect to receive from their children. Malay society is similar to the Western model, in that economic or other returns from children are not stressed. However, it is necessary to examine the parent–child relationship more closely in order to understand why the returns that *are* expected from children are expressed only in vague and tentative terms.

Uncertainty over the care children may give to their parents in old age is related to the conditional nature of the parent–child bond. Although it is believed that a child will remember and be grateful for the care he or she received when young, various factors could intervene to disrupt the parent–child relationship or to make return impossible. This stage of the family cycle will be discussed in detail later. At the time when the child is still young, a strong degree of certainty is expressed in the view that a parent who *fails* to care properly for a child, because of carelessness, cruelty, or because the child is being raised by someone else, will not be close to the child in adulthood. In old age, the parent will find the relationship with the child strained and unsatisfactory from both an emotional and financial point of view. Where there is a close and lasting bond, this is often explained by the care the junior party received while a child: 'I'm not close to my mother, because she did not raise me from small; I'm much closer to my grandmother, she's the one who raised me; I really love her. I can never forget her.'

In Java, the obligation of a child to care for parents in old age in return for the care he or she received when young, is apparently made quite explicit. Jay (1969: 76) writes that adoption involves a transfer of 'rights to material care from [the child] on his maturity, for which the receiving couple accepts ultimate responsibility for the child's upbringing'. According to Hildred Geertz (1961: 38): 'There is one primary concern of the parent and the adopting person, kinsman or otherwise: when the child is adult, whom will he support in their old age? The duty of the child to support his parent in old age is the aspect of the parent–child relationship which is most emphasized both in discussions of general relationships and in critical gossip.'

Jay's discussion of adoption in Java confirms that it is indeed *care* for the child, rather than consanguinity, or duties and rights defined by status alone, that is the key issue in the question of future old-age support. Geertz's analysis differs from two detailed studies of adoption in Malay society, which describe the range of factors that contribute to the decision to adopt a child. The studies do not find that old-age security is the primary factor and it is certainly not the only factor involved (Djamour, 1959: 93; McKinley, 1975). In contemporary Singapore, there are more ambiguities in the parent–child relationship than a simply stated obligation of the child to provide old-age support in return for the nurture received when young, and these will become evident as the later stages of the life cycle are examined.

Parents and Working Children

As children mature and begin to work, they give some money to their mothers. This is seen as payment for room and board, and as a gift made out of love for the parents. These senses are not clearly separated. The mother prefers to think of the services she performs for her family as her gift made out of love and concern for their welfare, a gift related to the mother's sense of the value of her labour power, which was described earlier. The mother would be offended if the money given by the children was clearly described as a wage, because that would devalue the kinship sentiment conveyed in her services when they are bestowed as gifts. The children also prefer to claim that they give money to their mother out of love and concern, and reject the idea that it is a wage since 'she is not our servant'. The ambiguity between the commodity exchange and the gift element is sustained by means of the budgetary practice. The children simply state they 'give money to mother' and do not specify further, while it is the mother who budgets for food and amenities and derives her personal income from any surplus.

There are occasions on which the commodity exchange aspect of the relationship is made explicit. These are usually occasions of anger and frustration, when children fail to give their mother sufficient cash and she either has to supplement the budget from her own resources or is unable to keep for herself a margin she considers acceptable. As a mother stated: 'It is only called "children giving mother", but it is not really ours. We have to give them food and pay for amenities. It is not as if we keep the money ourselves. The children don't even know if the money they give is enough for all the expenses. They say they give me money, but where does it go? Into their stomachs; I don't get any.'

The labour component may also be made explicit, as when a mother stated to her errant son: 'I wash your clothes, I cook your food, now I want my wage.' In one family studied, the father had made a rule in his house that working children living at home had to pay for their food, the utilities, and their mother's labour in washing clothes, and he claimed to be prepared to close his doors to any children who failed to contribute. This was regarded by neighbours as highly unusual, but also wise, since many of them had experienced the frustrations of providing food and services for working children who gave nothing in return. The reluctance to make the commodity-exchange element explicit, stems from the preference of both parents and children to see the relationship as one built on kinship sentiment, in which the flow of goods, services, and cash is motivated by care and concern and not reducible to market terms.

The example of baby-sitting, a service provided by mothers to working daughters, further illustrates the tension between the commodity and the gift elements in parent–child exchanges. Grandmothers who stay at home to care for the children of their working daughters or sons see their service as a gift made out of their love and concern (*kasihan*). They are aware that the money they receive for baby-sitting is less than the current market rate, but agree to do the task because of their concern for the daughter who has to work to boost her family's budget, and concern for the grandchild who may not be cared for properly by outsiders. In families where the husband's income is low, the wife's ability to work during the first ten years of the marriage enables the couple to acquire essential consumer goods and accumulate some savings for the children's education in the long term. The wages available to women with low education, as factory workers or domestic servants, would barely cover bus fare and baby-sitting if the latter were paid for at the commercial rate. The availability of the subsidized baby-sitting services of a grandmother is thus a highly valued asset.

Grandmothers are usually paid $80–100 per month for baby-sitting, which is less than the usual rate of $150–250. Grandmothers rarely complain explicitly about the 'low pay', since that would devalue the element of *kasihan* or voluntary giving out of concern which is the spirit in which they see their actions. Likewise children, when giving the monthly money to the grandmother, do not call it a wage or baby-sitting fee, except jokingly, because that would be to treat the mother as a paid servant. The vagueness is sometimes a conscious subterfuge: 'We make sure we don't give my mother exactly the same amount of money each month for looking after our child. If we did that, it would be like a wage. Sometimes we give more, sometimes less. We say, "This is for you, for your pocket-money; if you need more, we can give you more."' Yet grandmothers who look after grandchildren always receive more than grandmothers who do not, as the usual token that a child gives an elderly parent for pocket-money is $10–30. This reflects the underlying, though unspoken, commodity-exchange element.

While grandmothers do not complain about their wage, they often complain about tiredness. They can frequently be heard to comment that 'old people now have really become the children's slaves'. There is some rivalry among siblings over access to the subsidized baby-sitting service of grandmothers, and those not enjoying the service usually express concern for the old lady's health:

My siblings all put their children at my mother's house, I don't. I ask my neighbour to take care of my child. Because the old should have a chance to rest, not be tired out with child care.

Grandmothers are often explicit about the wages they have forgone:

I used to work six days a week. Then my daughter wanted to work. I felt sorry for her (*kasihan*); her husband does not earn much. So now I only work three days and my daughter works three days too, while I look after the child.

Other grandparents are less content with the relationship:

Children now are too clever. They want more money so they give other people trouble. My wife was getting $10 a day in her job; now they make her look after the child, and how much do they give her?

Some grandmothers refuse the role:

I have five children, all are working in good jobs. I have ten grandchildren. But I don't look after them. I prefer to work [as a canteen cleaner]; I can't be bothered to look after grandchildren. They pay other people to do it. If I stay at home, I wouldn't get any pension.

The desire for an independent cash income without the ambiguity of money given by husband or children was discussed earlier. In a highly commercialized economy, the power and sense of generosity that accompanies gifts of cash, outweighs that of gifts of labour service. One grandmother, who was working as a domestic servant in order to be able to help her son financially and to express *kasihan* for her grandchildren in material gifts, preferred this arrangement to the alternative of staying at home to look after the grandchildren, which would have enabled her daughter-in-law to work. Another grandmother took the opposite approach, and claimed that she did not want to work because if she had her own income, then her children would not give her any money so she would lose out on a possible source of 'free' income.

Parents' sense of generosity, of continuing to give to their children even after the latter begin work, is shown not only by their willingness to provide subsidized food and labour services, but in their reluctance to demand or accept money from their children. This could be called the 'gift of not taking' and it is clearly underwritten by a sense of legitimate entitlement based on the values of commodity-exchange. If seen as commodities, the goods and services provided by parents to working children in the home are being 'underpriced'; 'given away' cheaply or actually free of charge.

My daughter gives her mother a bit of money, but it doesn't even cover her expenses for food and amenities. Wait until she has her own house, then she'll see how much these things cost. They should give us money, but then the parents don't want to take it, they say never mind. We feel sorry for the children (*kasihan*); we don't like to take their money; because we don't think about becoming rich. If we have enough, why take money from others?

In this case, the father earned a high salary and was not in need of the daughter's contribution. The mother who accepted the daughter's money was not using it but saving it with the intention of exercising *kasihan* herself by making gifts of goods and money to the daughter when she set up her own home. From the daughter's point of view, she was just 'giving mother money' and what the mother did with it was not her concern.

No clear variations were evident in the way this relationship is seen by Malays of different social and educational backgrounds, although there were differences in the actual amounts of money and other services transferred. Some parents who are well off, refuse all contributions from children, saying they can afford to provide food and services in the house, so children who wish to remain in the parental home can stay free of charge. When claiming that goods and services are provided freely in the household, or claiming that he or she does not like to calculate the balance of exchange, the parent is claiming to make a gift to the child. The calculation of values on the principles of commodity-exchange is integral to this form of gift, since in the commercial economy only items of recognized commercial value count as gifts. In the delicate way these exchanges are negotiated in the course of daily life, the commodification of goods and services in the household enhances the idea of the gift. However, this is so only where kinship sentiment is strong and the individuals involved wish to sustain it. Where kinship sentiment is weak or absent, or in situations of anger which can occur in any household on occasions, the high level of commodification provides the ready means to calculate unevenness in the balance of exchange. If the calculation is made open and explicit, it negates kinship sentiment built on the willingness to give.

The householding relationships described above reflect the balance of commodity-exchange between parents and working children in the current household budget. There is a second set of terms used to describe the transfer of money from child to parent, which is related to questions of power and responsibility in the extended household cycle. Some children have come to regard the money they give their parents as 'help' (*bantuan*) or as their 'responsibility' (*tanggungjawab*). This is not responsibility in the sense that their wages are *owed* to a joint family economy, or *owed* to their parents in return for nurture received when young, but responsibility the child as an individual has voluntarily assumed. Desire to take on this responsibility arises from concern for parents' and siblings' welfare, and it may also be related to the child's gratitude for the parents' care and sacrifices. It is a responsibility motivated by kinship sentiment, and not imposed by a sense of definite obligation or debt.

The superiority children in Singapore can assume as the providers of financial support to the family, can shade into terms of patronage. This is evident when children begin to describe their financial contributions to their parents as gifts made out of pity and concern (*kasihan*): 'I sometimes help my father pay the utility bills because I pity him (*kasihan*), he has so many things to pay.' Says another, 'I always try to give my mother money because I pity her (*kasihan*) now she is not working. Maybe she won't have enough money if she wants to go out and buy something.' It was mentioned earlier that the term *kasihan* has strong connotations of power and condescension. For parents to see their relationship to their children as both responsible and generous, has deep roots in Malay culture, as shown above. For children now to begin to express their relationship to their parents in terms of responsibility and pity, suggests a transformation in the relationship between the generations in which children have begun to take on a quasi-parental role. This cultural change is best understood by comparing the economic and social conditions prevailing in Singapore with those of the rural Malay and Javanese world.

A major difference between the pattern of rural Malay life and that of contemporary Singapore Malays is a demographic one which concerns the rising age at marriage, particularly for girls. The postponement of marriage creates a new phase in the family cycle in which children are economically productive, but unmarried and living in the parental home. Both male and female children are economically productive in Singapore, since they begin to work as wage-earners as soon as they leave school. In the rural Malay economy, children become major producers only when they are given their own land and capital, or begin to farm as share-croppers or tenants on their own. This is a transition that occurs upon marriage or soon after, which is also the point at which they leave the parental household. In the urban wage economy of Singapore, the presence of unmarried working children in the Malay household could potentially signal a reversal in the downward flow of wealth from parent to child which has generally been characteristic of the rural Malay household economy. Singapore Malay children could become substantial contributors to their *parents*' livelihood and store of wealth and savings. Whether this occurs depends on how this life-cycle phase is managed culturally. The following sections discuss in detail the financial contributions to the parents made by unmarried boys and girls, and then turn to the question of authority in the relationship between parents and children in Singapore.

Parents and Unmarried Working Daughters

The demographic and economic changes noted above have been most dramatic in the case of girls. It has long been the pattern in the Malay and Javanese rural hinterland for girls to marry at, or shortly after, puberty. In Singapore, the median age at marriage of Malay girls has risen from 18 years in 1946-50 to 24 in 1982 (Chang *et al.*, 1980: 65; *Economic and Social Statistics of Singapore*, 1982: 23). The participation of Malay girls in the Singapore labour force has also risen very rapidly. In 1957, only 6 per cent of all Malay females were in the formal labour force, and this had risen to 42 per cent by 1979 (Wahidah Jalil, 1981: 83). The greatest increase has been in the labour-force participation of unmarried girls which began in the mid-1970s with the opening of employment opportunities in the electronics assembly plants of multinationals. In 1980, 68 per cent of Malay girls aged 15-24 were working, while a further 17 per cent were in school, leaving only about 15 per cent of the age group at home (Census 1980: III.20, II.43, and IV.41).

In both rural and urban contexts, unmarried Malay girls who are not employed outside the home provide unpaid domestic services to the household. These services are a contribution to the current reproduction of the household, and also an indirect contribution to the mother's personal economy, as the mother is relieved of domestic duties and able to seek an income outside the home. In the rural economy, husband and wife are able to concentrate on increasing their rice production and accumulating assets during the stage when daughters are old enough to help at home. White (1976) has documented the domestic-labour contributions of female children in Java and argues that girls are of considerable economic value to their families. The crucial point, however, and one not recognized by White, is that the girls' labour is not conceptualized as work nor recognized as a contribution to the household. The household labour of girls tends to be viewed as 'helping mother' or 'learning' the adult female role (McLellan, 1985). Unlike their mothers, the girls do not control the budgetary margins as a potential source of personal cash or as a substitute for a wage which could compensate them for their domestic labour. They are not entitled to a share of their mothers' income, which the daughters' labour actually helps to make possible. When girls marry, their parents give them agricultural capital, household goods, or part of the customary marriage payment provided by the groom, but this is seen as a gift from the parents and not as an entitlement. It is not seen as a share in a joint estate to which the daughters have contributed through their domestic labour.

Of all the members of a Malay household, it is unmarried, non-earning girls who place their domestic labour most freely and fully at the disposal of all other household members. Older women looking back on their youth always note how they felt 'stupid' or ignorant and just did what they were told. In Singapore, the opportunities for unmarried girls to earn a cash income have changed this situation. Not only are girls better educated than their mothers, they can make *explicit* their contribution to the household, and to their mothers, by making it in the commodified form of cash or consumer goods. In this form, their contribution has become a gift which they can bestow or withhold as they choose, since they are paid directly by their employers. This has brought a new tension to the relationship between mother and daughter. No longer can mothers assume that they will enjoy the total benefit of the girls' labour services in the household. They now have to negotiate for a share of the daughters' wages which they do not control and upon which there are competing demands. There is no existing cultural legitimization for the view that the product of the girls' labour belongs by right to the mother. Once the girls' labour is clearly seen *as* labour–that is, as soon as they work for wages outside the home–Malay girls have attained the right to enjoy the product of their own labour that has always been enjoyed by adult Malay men and women, and, it will be shown, also by boys.

The arrangement in most Singapore Malay households in the 1980s is that girls give their mothers a fixed monthly sum, usually a third of their pay. Typically, a factory girl with a take-home pay of $300 gives her mother $100, spends $100 on transport and expenses, and saves $100 for consumer purchases for herself or the household. A better-educated girl earning $900 gives her mother about $300, the proportion remaining roughly the same whatever the income. The girls see this transfer in the range of ways described earlier. It is given as payment for food and services, made explicit by a daughter who said, 'No, I don't do any housework at home, I give my mother money, that's enough.' It is also given out of a sense of responsibility and desire to help the family, and it is given as a gift, made out of a sense of love and concern for the parents. The amount to be given is negotiated, though the stronger voice rests with the girl who has the cash in her hands. The mother nags if she feels the sum given is insufficient. The basis of sufficiency is related to the cost of food and the margin the mother expects for herself, taking into account the daughter's wages, and the margin the girl expects to retain for her own consumption and saving. The mother may well succeed in accumulating a surplus from the child's income, but she is likely to downplay this, commenting on the high cost of food and utilities. A girl who gives nothing or little will be criticized by her mother and onlookers with comments focusing on the commodity-exchange element–'She thinks she can eat free.' A girl who contributes a large part of her pay to help finance the maintenance of the whole family and the education of siblings is praised. She is said to be good and responsible. A girl who gives generously though her mother does not really need the cash is said to really love her mother, and the mother is lucky to have such a daughter. However, it is felt that the mother should not try to extract more money from her daughter than she could reasonably be said to need, since it is the girl who worked for the money and it belongs to the girl herself.

It is the emotional closeness of the girl to her family and particularly to her mother that is felt to be the source of the girl's willingness to help shoulder the burden of family maintenance. It is also the source of the girl's decision to be generous, which may allow the mother to accumulate some funds. Mother and daughter sometimes decide to prolong the period during which the girl can help her family or provision her mother with some savings and a few luxuries. The mother has an obvious interest in this postponement, and may declare it explicitly: 'I tell my daughter don't get married too soon, give mother a good, full share first.' Many girls are happy to postpone marriage for this reason: 'I say it's better not to marry; it's better to work yourself, then you can give money to your mother. Once you marry, you don't know whether or not you will be able to give to your mother. Maybe there will be no one to help her.' Girls may also wish to prolong the period of personal freedom and affluence which, unlike their mothers, they have gained before marriage. A mother commented that her daughter 'says she hasn't had enough enjoyment of being single yet. She likes to go out. She has her own money, why would she want to marry?'

Several cases were encountered where girls remained unmarried into their forties in order to work to support their parents and siblings. Social pressures are such that most girls do marry, though, as noted, the age at marriage is rising. A girl who never married would be considered eccentric or would be pitied because of the loneliness she would face in old age with no children of her own. The flow of cash from daughter to mother inevitably drops as the daughter marries and needs resources to establish a new home. If the daughter stops working, the flow may stop altogether or be reduced to a token, saved out of the housekeeping or specially requested from the husband. Some girls have ensured, before agreeing to marry, that their husbands will allow them to continue to work to 'help mother'. Once children are born, if these are looked after by the grandmother, the relationship changes again as new calculations and possible tensions are brought into the exchange, as described earlier.

Ideally, mother and daughter see the relationship between them as a voluntary and loving one. The daughter gives out of concern and love for her mother; the mother herself exercises *kasihan* by not demanding too much, allowing the girl to enjoy her own wages. There is no precedent in the Malay cultural heritage for a mother to accumulate cash from her daughter's labour (though, as noted, rural mothers did so indirectly). The changed economic circumstances that leave daughters relatively affluent have been managed culturally in the idiom of kinship sentiment which *may* enable the mother to accumulate wealth from the daughter's labour, but which does not give her any clear right to do so. The mother's accumulation is ultimately the result of the daughter's gift, and though the mother may have high hopes, excessive nagging can only undermine the kinship sentiment upon which the flow of cash depends. Some daughters were heard to complain that 'my mother doesn't really love me, she only wants money'. One girl who had not been raised by her own mother described their poor relationship:

I am not close to my mother. She did not raise me when I was young. She does not really love me. She has favourites among the children. If my brother doesn't give her money she doesn't say anything, but if I don't, she makes a lot of noise. Now I'm married I don't give her anything. I heard she's really having problems now I have moved out. Without money she's in trouble, she is sick–I mean really upset (*sakit hati*).

Clearly not all relationships are negotiated successfully and the currencies of loving care and of cash are closely and precariously intertwined. The flow of material gifts upwards from child to parent is described by the same terms, generosity and love, which are used to describe the downward flow of nurture and material wealth from parent to child in the rural economy. Where daughters are concerned about their family's welfare and are generous with their cash, there can be a reversal of the flow of wealth at this stage of the family life cycle in Singapore, but this has not entailed a radical restructuring of the family economic system to one based on the rights of parents to children's labour, as in a joint family economy.

McLellan (1985) has made a comparable analysis of the relationship between mothers and working daughters in rural North Peninsular Malaysia. She heard a mother express her dissatisfaction that, so far, her working daughters had not yet 'made a return for the milk given them'. McLellan interprets this as an indication that children have a clear obligation to give money to their parents, and she goes so far as to suggest that the expectation of material return is the reason why Malays have children. In Singapore, though this expectation is present, especially now that unmarried children can earn wages, it is a source of negotiation, anxiety, and often bitterness precisely because the obligation is conditional and not clearly defined, as it would be in a joint family economy. McLellan reports that mothers resort to witchcraft to ensure that their daughters hand over a share of their pay. An alternative interpretation of the observations she made, concurrent with the analysis of family relations presented here for Singapore, is that mothers resort to witchcraft because it is a mechanism for changing the *mind* of the daughter, making her feel closer and more loving towards her mother. If the mother felt she had a clear right to her daughter's wage, she would say so directly and state the grounds, supported by custom and religion, but in fact these grounds are absent. A tough stand by the mother would be likely to weaken the very emotional bonds on which she must rely for both social and material returns from her children.

It has been established here that the flow of gifts may in some circumstances flow predominantly from child to parent, but there is no precedent in the Malay or Javanese rural cultural heritage for unmarried girls to shoulder the burden of supplying the basic consumption needs of their parents and siblings. While mothers seem to have few qualms about this new development, saying times are hard and they are grateful for their daughters' efforts, fathers show more reluctance to make economic demands on their daughters. No fathers are at all proud of the fact that their daughters contribute a large part of the family budget, and most fathers try to distance themselves from these transactions. They stress that they never demand nor receive any money from their daughters, though they acknowledge that the daughter may give to her mother. The fathers assert that if such a transfer is made, it is voluntary, and not really necessary since they have already provided for the family's requirements. They often profess an exaggerated disinterest in this transfer: 'I would not like to tell my children, girls or boys, that they have to give me $100 or $200. If they want to give some money, they can give it to their mother. Maybe they do give to her, but I never ask how much. I don't ask how much they earn. I never ask for any money from them. If they want to give to their mother, that's up to them.'

Fathers also like to think that their daughters love them, and may express this by saying 'she always wants to give me money' but they in turn express their love and generosity by turning back these gifts: 'I tell her to keep her money for herself, she'll need it for her own future.' If they do accept, they stress the voluntary, gift aspect: 'Well, she wants to give, it is not as if I asked for it.' The impropriety fathers feel in accepting gifts or contributions from children, boys or girls, will be discussed further on pp. 56-9.

Parents and Unmarried Working Sons

Boys also give money to their mothers rather than their fathers, but the parents' expectations of boys are rather less than the mothers' expectations of their daughters. Like their sisters, some boys do take on responsibility for their parents and siblings in cases of financial need, and contribute a large part of their pay to their mothers for current consumption and the education of younger siblings. As in the case of girls, this is considered praiseworthy, but it is not expected as a matter of course. Where sons spend most of their wages on themselves despite family need, this is not seen as unusual or at all surprising, though it may be frustrating and disappointing to the parents. It is, after all, the boy who works and his money is his, while the ultimate responsibility for maintaining the parents and siblings rests on the father. The few fathers who do accept or demand money from unmarried sons again downplay their need for the money. Some fathers state that they accept money only in order to train their sons to be responsible husbands, good managers of money, and thoughtful of their mother in case they should one day need to support her after the father's death.

Mothers expect less from their sons for two reasons. First, boys are felt to be less close emotionally to their mother, so that the bond of kinship sentiment which channels the girl's wages to the mother is expected to be weaker. A boy who does give his mother a large share of his wage occasions the comment 'he really loves his mother'. Secondly, while girls are expected to spend most of their free time at home, boys are expected to have peer-group activities and expensive tastes in clothes, entertainment and cigarettes, all of which require substantial amounts of cash. Parents accept and condone this pattern, and comment only that 'boys will be boys'. Fathers recall the similar period they enjoyed in their own youth:

My children don't give us any money. They profit, we lose. Because whenever I see these bachelors, I think, 'Let them enjoy themselves like we did before, go out and see friends, have a good time.' If they go out with friends they need their own money, they can't expect their friends to pay for them all the time. So I let them be.

Fathers often claim to have given more money to their own mothers in their bachelor days, usually half their pay, but they relate their sons' failure to contribute to the higher cost of goods and the greater range of activities that boys in the 1980s can enjoy. They do not feel there has been a radical change of mores. In the past, too, some boys gave nothing, others gave if they chose, and parents generally did not demand money from their sons. It was the son's decision whether or not to contribute. There has been only a small rise in the age at marriage for Malay boys in Singapore, as the majority both in 1957 and in 1970 married between the ages of 25 and 29 years (Chang *et al.*, 1980: 51). For boys, there has been only a slight prolonging of the period of adolescence, but not a radical change in their economic position in the family equivalent to that experienced by girls.

Adolescence in Singapore has brought added dangers to Malay boys in the form of drug abuse. Malay boys were particularly affected by a heroin epidemic which was at its peak from 1976 to 1978. During that period, Malays represented 30-40 per cent of those arrested and placed on supervision or sent for rehabilitation (SANA *Annual Reports*, 1979-82). A total of 3 per cent of the 15-24 age group, comprising all ethnic groups in Singapore, was estimated to be abusing heroin in 1977, and an even higher proportion of Malays were affected (McGlothlin, 1980: 4). It could be that the financial irresponsibility and peer-group involvement Malay parents expect from their sons, accounts in part for the disproportionate Malay representation among the drug abusers in Singapore. The approach many Malay parents take is that excessive nagging for money will drive boys to escape from the home to the company of their peers, or, worse still, to steal for money. So the parents do not demand money even to cover the cost of the boy's food, and in many cases give them extra spending money although the boys are working: 'I say to him, "If you don't give me any money, never mind; I can find money, so long as you don't turn bad."'

The life-cycle stage of adolescence is an ambiguous one for Malay boys. From the age of circumcision (about 10 years), they feel they are 'already big' and under Muslim law they take on adult responsibility for their own prayers and religious duties. They begin to resent parental authority, saying, 'Why do you want to control me, I'm grown up,' and parents do not feel they can or should control their sons too much. There are often strained relations between father and son at this stage which are handled by avoidance or by the sons' absence from the house, visiting friends or kin. Though big enough to manage their own time and relationships, boys are not expected to be financially responsible for themselves or to make regular contributions to the household. The transformation to a state of real adulthood and financial responsibility is expected to occur around the time of marriage. The boy is expected to save the substantial sum he must give to the bride's family, and many boys start working regularly or doing overtime to achieve this. Upon marriage, they are responsible for their wife and children, though it was noted earlier that many boys take some time to adjust their life-style and spending pattern to this new role.

Rural ethnographies show a very similar pattern in the relationship between boys and their families. Boys are not expected to contribute regularly to their parents' economy either in cash or by work on the land. They keep their earnings to themselves even while eating and sleeping in the parental home. They are expected to work seriously and to be financially responsible upon marriage, though the benefit of their labour then goes to their new nuclear unit and not to the parents.

In rural Java, according to Jay (1969: 35, 68, 69, 138), boys work at jobs they largely find for themselves; there is little conceptual stress on the child's labour as a contribution to the household and children are expected to spend their earnings on themselves and their peers, though they are praised if they hand them to their mother. After marriage, parents begin to pay their children for their labour, to enable them to save and accumulate capital to set up a viable farm of their own. Though sons and, especially, sons-in-law do some unpaid services for the father, the parents attempt to hasten the young couple's separation into a new independent household, and do not try to prolong the period of access to their labour.

There are several sources of variation in the nature and extent of labour performed by children in Java. White's (1976) study of the productivity of children's labour in Central Java establishes that both boys and girls work long hours in the rural sector. Hart (1986: 129), writing on a coastal community in Central Java, notes a clear relationship between the income-earning activities of children, particularly girls, and the asset-status of their household, with girls aged 10-15 from poor or landless households engaging extensively in rural wage-work. Hardjono (1987: 196), describing a West Javanese village, finds that children under 15 years of age make very little contribution to farm work, labour being provided principally by husband and wife, but she provides little data on children's off-farm, waged, or domestic labour. Unfortunately, none of these authors gives much information on the actual material exchanges between parents and children, or the conceptual terms governing the exchange in either the short or long term. Unless it is known, for example, whether or not the small boys gathering fodder for a neighbour's livestock hand over all or a percentage of their earnings to their parents, the productivity of their labour does not prove that they are productive *for their parents*.

Bailey, who compared three types of rural economy in Malaysia, noted a similar pattern in each. In the rice economy, boys help at periods of peak labour demand in the annual cycle but they keep other earnings to themselves (Bailey, 1983: 76). In the fishing economy, unmarried boys give most of their fishing money to their mothers but when a boy marries, any 'subsequent loan from his parents may be considered not so much as a matter of inheritance or cross-generational assistance as recognition of past contributions to the family which in this manner are being repaid' (Bailey, 1983: 182, 103). Here the transfer of money from son to mother is a form in which the son saves and accumulates the product of his own labour for his own future use. There is no suggestion that the family as a unit, or the parents, have any right to his labour, nor that he owes them a payment for his upbringing. It is in areas where rubber production is a principal source of income, that boys are least involved in contributions to the household economy, according to Bailey (1983: 139):

There seems to be a period during which male adolescent Malays become rather frivolous and irresponsible, when they prefer to 'hang out' with their contemporaries. This is ... most common among offspring of relatively wealthy families.... Social prestige accrues to a family which can afford to remove from remunerative employment these potentially productive members.

Wilder (1970) has also stressed the lack of economic contribution of unmarried boys in Malaysia. He relates this both to the prestige of the family in keeping the boys idle, and to the enforcement of authority over boys during the ambiguous period when they are 'big' but not yet married. In Singapore, there is no evidence that idle boys bring any prestige to Malay families, since such boys are likely to be in some sort of trouble and to spend their time with undesirable friends. Prestige accrues instead to families whose boys have studied well, are serious, hardworking, and both generous and responsible towards their families. Middle-class families are more likely to achieve that ideal, and its achievement is an important indication of class status, a subject to be discussed in Chapter 8. In general, more is expected of the ideal Malay youth in Singapore in the 1980s than of his rural counterpart, but these greater expectations have not been codified into a new set of rights and duties in which children *owe* their parents a share of their labour. Emotional ties and general religious and moral training are relied upon to produce upright sons who will voluntarily give money to their parents. It is now said that parents have only themselves to blame if sons turn out to be bad and irresponsible, though the means for enforcing parental authority over children have become weaker than ever.

Authority and the Flow of Giving

The economic conditions of Singapore, particularly the relative affluence of the young, have had a major impact on relationships of authority and dependence between parents and children. The change was succinctly stated by one parent: 'Now it is they who give us food, not we who give them food; we can't say anything.'

The significance of giving food in setting up a relationship of authority has been extensively treated in the Malay and Javanese ethnographies. At its simplest, 'to give food' means 'to have authority'. When complaining about a neighbour who has disciplined a child not his own, or a factory supervisor who is heavy handed, Malays comment that 'He should not do that; who is he anyway? It is not as if he had given food.' A similar observation was made by Djamour (1959: 45).

The authority that Malay parents have over their children is based on their position as the providers of life, care, food, and property. McKinley (1975: 289) argues that it is not possible for the flow of transfers in this relationship to be reversed. It is not possible to move from 'being the recipient to being the giver in this relationship. Since it is one's very life, sustained by parental care and nurturance, that one has received in this relationship, anything that one might give in return would be beside the point of the relationship itself.'

It is true that the gift of life, like any gift, is irreversible. The gift is returned, in as near equivalent a form as possible, in the care given to the old at the end of their lives, as will be described later. Nevertheless, the nature of parenthood, which is traditionally defined by a downward flow of sustenance and material goods from parents to children, has been significantly altered in the present economic conditions in Singapore. Children in Singapore can be major contributors to the household food budget, providing for themselves, their parents, and siblings, and even giving their mother a margin for accumulation. This they do while they are still unmarried, living in the parental home, and long before the parents are old or feel ready to accept a subordinate role as recipients of food and care provided by the children.

Parental authority is potentially undermined by the economic contributions of working children to the current household budget. This explains why the parents, and especially the father, feel *ashamed* to take the children's money. The father accepts as little money as possible from the children, downplays it, diverts the money to the mother, or regards the transaction as 'training'. The model of parenthood defined by providing for children is sustained by the father's refusal to accept or demand financial aid which the children are potentially able to give. When parents refuse contributions, or assert that they do not demand contributions, they say they are showing *kasihan*, or generosity and pity for their children. This refusal to demand or accept money from children counts as a significant gift to the children only if the children have an unspoken obligation to contribute money to their parents. The suggestion is made that the children 'ought to give but they need it themselves, or are spendthrift, or we can manage without it'. The feeling that the children ought to give their parents money to cover their own food, to 'help' their parents or siblings who are in need, or to allow the mother some accumulation, is accentuated in Singapore. Rural Malay children were not expected to contribute much to their households since, for girls, the life-cycle stage of unmarried but earning did not exist at all, and for boys it was given little significance. Though parents do not claim a share of the child's pay as of right, it is clear that contributions from children have come to be expected. This is made clear in the praise of those who give, and even more so in the grumblings and frustration directed at those who do not.

At the same time as financial contributions from children have come to be expected, parental dignity, defined by providing for children, has been retained by stressing that any contributions from children are voluntary gifts. Gifts a child makes to a parent out of love are less threatening to parental authority. Children are aware of the potential embarrassment to their parents of becoming the recipients of aid, or even charity, from their unmarried children. Those children who express *kasihan* for their parents do not do so in their parents' presence, especially not in front of their father. Some children are consciously engaging in a subterfuge, whereby aid has to be described as gifts, cigarette money, or else directed through the mother in order to make it less damaging to parental pride. 'Giving money to mother' is a most ambiguous act. The final destination of this money, whether it is for food, the mother's wage, a gift, or to pay a sibling's school fees, is left totally unclear. This appears to be a necessary vagueness in a system of expectations and obligations, gifts and counter-gifts, where the crucial element is that no exact reckoning could or should be made. Where kinship sentiment is positively valued and current parent-child relations are reasonably harmonious, the feeling any party might have that the balance of exchanges is to their individual detriment is not directly expressed. Instead, the individual claims that he or she does not calculate, or is happy to make gifts.

These are the cultural inventions by means of which the shift in power and affluence between the generations has been managed by Singapore Malays. It is not to be assumed that the practices described harmonize into a clearly defined and approved code of conduct. There are constant strains, tensions, and negotiations dealt with by constant invention. There are mothers who complain that their daughters have grown arrogant 'now she gives me a wage of $200, she thinks she is so important'. There are children who use the monthly 'pay' they give the mother to demand privileges in the home, and threaten to withdraw the money if the mother nags too much. Children with education sometimes feel they 'know more' than their parents, and mothers complain that the young have become insolent. At worst, children can come to scorn their parents for their low education and occupational status: 'You are just a road sweeper. What do you know?' Generally, though, the shift in authority has been less obvious and less brutal. Formal respect for parents is retained and parental dignity preserved, at least in appearance, by the cultural strategies described.

One area that writers on other Malay societies have stressed as an important sign of parental authority over children, is the submission of children to their parents' choice of marriage partner (Wilder, 1970: 230; Hildred Geertz, 1961: 56). The data on contemporary marriages obtained in the course of this research showed that most parents no longer involve themselves in arranging their children's marriages or attempt to exert authority over their children in this respect. They relate their attitude of non-interference in part to the children's economic independence. In the rural Malay or Javanese setting, a boy would usually be dependent on his parents for the provision of the cash sum he must give to the prospective bride, and for the land or other agricultural capital needed to establish the new household. In Singapore, where the boy earns his money himself, parents say, 'Of course the boy has to find a wife and pay the money himself; he's the one who wants a wife, not us.' In the absence of financial dependence, there is no obvious way within the Malay cultural framework that authority over children's marriages could be enforced. Singapore Malay parents can exert only moral pressure and persuasion if they disapprove of a marriage choice, and even this can backfire if excessive parental interference alienates the boy and weakens the emotional tie that is so highly valued.

In the case of girls, parents are concerned to exercise direct authority in order to ensure that their daughters marry as virgins. In Singapore, it seems that the benefits to the parents of having working daughters outweigh the potential dangers of the daughters' absence from home. Parents differ widely in the amount of control they exercise over girls after work, and there is no doubt that the girls' financial independence has made parental control more difficult. The conduct of daughters depends on individual character and training, especially religious training, a subject upon which many parents in the 1980s place strong emphasis. In Singapore, the deepest change in the relationship between parents and unmarried children is not in the loss of direct or petty authority over their daily lives or even their choice of marriage partner, but in the potential undermining of the whole basis of parenthood defined by the downward flow of sustenance and gifts from parent to child.

Apart from daily sustenance during the years while they grow up, the principal gift that rural Malay and Javanese parents give to their children is the agricultural capital with which to earn an independent livelihood. This is the gift that enables the children to become economically independent and hence fully adult. The gift of property and the transition to adulthood take place on marriage. Young Malays in Singapore achieve their independence in a series of stages, making the whole transition more ambiguous for both parent and child. More importantly, the children's financial independence no longer comes as a tangible parental gift in the form of land, but is an independence the children find for themselves as they begin to work for wages. Though the gift idiom has become so prominent in the exchanges of daily life in the Singapore Malay household, the major material gift rural parents give their children is absent or less relevant in Singapore. This has implications for the ways in which parents are able to provide for the future of the children they have raised to adulthood (see pp. 78-83). It also has implications for the relationship between parents and children in old age. In the rural economy, the parents' possession of material assets which they can pass on to their children or withhold at will, gives them a basis of security in old age. Old-age security too has been undermined in Singapore, as will be shown through analysis of the third phase in the family life cycle.

Parents and Children: Old Age

Reference has already been made to the third stage in the family life cycle, when the old become dependent. It was noted that the parents likened their old age to a period of reverse infancy, when the children they once nurtured can in turn nurture them. Others have noted this reversal. Among Javanese villagers, Jay (1969: 108) noted,

> The parent becomes the dependent and therefore, although still treated with deferential courtesy, begins to accept direction from the child instead of giving it.... [They] viewed the end relationship as good and beautiful, and parents (and parental surrogates) looked forward to such an eventual relationship with their mature children as the source of security in old age.

In Singapore, the relationship seems to be marked by two sources of ambivalence. The first is that parents desire to retain their independence for as long as possible. Fathers remain unwilling to accept food from children until they can no longer work: 'It doesn't feel nice, to sit down and take the children's money. I prefer to work myself.' The parents are acutely aware of their loss of direct authority in the household as they become the recipients of food: 'We don't say anything, it is their household, we're only hitching along; they are feeding us, not us feeding them.' Hildred Geertz (1961: 34) records that in urban Java, too, old people who lived with their grown children 'never asked their child for anything, simply accepted what was given them, and never interfered in their child's economic arrangements, never gave advice or even know much about the family budget'.

The powerlessness and peripherality of the old people in the child's household is an incentive for the old to continue to maintain a separate dwelling and an independent income for as long as they are able. The 1980 census shows that 42 per cent of Malay males were economically active at the age of 60, and 30 per cent were still active over the age of 65 (Census 1980: IV.41, II.26), though the official retirement age was 55 years. There is more than a cultural preference for old-age independence involved, however, and the second source of ambivalence concerns the ability and willingness of the children to care for their parents. The tentative, conditional way parents refer to future support from their children was described earlier. Though parents may have no alternative but to rely on children, this dependence is expected to be problematic. The dilemma is captured in the Malay saying, 'To depend on one's own child is blindness in one eye; to depend on a stranger is blindness in both eyes' (Winstedt, 1950: 52).

The children's ability to care for their parents is in doubt because, with low education and low pay, most children have to struggle to meet the current needs of their own households for rent, food, and education (see Chapter 7). This is recognized by their parents, and the term parents use to express their concern for their children's financial difficulties is *kasihan*: 'Yes, children should look after their parents. But we also have to think. We feel sorry for our children. How much can they give? Even $10 is a lot to them. Their own children need school books and they need the money themselves.'

Here parents are making the gift of 'not demanding' the financial support in old age which they feel is really their due. Parents who do make heavy demands on their children, or who fail to work to support themselves while they are physically able to do so, are censured by other Malay adults: 'Why does he just want to sit back and depend on his children? He knows the children have their own problems.'

The children's willingness to provide for their parents is also in doubt. Deep insecurity over this is related to the conditional nature of the parent-child bond discussed earlier. It is expected that the care given to a child will create a lasting bond and provide some security for the parent in old age, but this is subject to three factors. First, how close was the bond established? If a child was raised in part by someone else, this can lead to divided loyalties and a lack of closeness to the parents. Secondly, there is the child's own individuality. It was noted earlier that some children are thought to be mystically unmatched to their parents through their unique personality, which is the gift of God. Others may, through the course of events or through the friends they keep, just drift away or turn out badly. Their relationship to their parents in old age can therefore not be guaranteed; 'We don't know yet whether or not he will remember or love his parents; I hope my child will feel pity for me when I'm old, but it's up to him.' The third factor is the child's emotional loyalty to his or her spouse. This is almost inevitably expected to be disruptive of the parent-child bond. It can turn a good son or daughter away from the parents. As one mother said in anger: 'What is a child? What is a grandchild? As soon as they get married, it is finished; there is nothing left. She listens only to her husband.'

The intervention of the son- or daughter-in-law is the principal reason why parents planning for old age say they cannot depend on their children. Not only do they not like to be dependent, but economic factors may make the child unable to provide and the way marital and parental relationships are organized in Malay society makes the child *undependable*.

The common arrangement for old age is for the elderly couple to live alone for as long as they are able. The father usually continues to work and covers at least his own expenses. The mother receives small contributions of $10-30 from each of her children and some of her grandchildren, which she uses for food. A stem household is formed when one child and spouse decide to stay in the parental home, or a widowed parent moves in with a child. This child maintains the parent on the basis that cooked food in the household is shared, and the contributions from other siblings become the old folk's pocket-money. It is very rare for elderly Malays, married or single, to be abandoned and left destitute by their children. Despite the insecurities expressed, the filial bond in fact turns out to be a strong one. In the few cases of recalcitrant children, or old people who have no children of their own and have not successfully developed a lasting parent–child bond through adoption, these unfortunates are taken in by other Muslim families on the basis of charity or adoptive kinship.

The most common dread expressed by old people is that though their children may take care of them, they or their spouses may do so unwillingly. Old people fear overhearing grumblings and complaints which make them feel uncomfortable and unwanted. They express the desire to die quickly so that they will not cause trouble or need to be nursed. Elderly parents prefer to stay with daughters since it is not expected that a daughter-in-law will be willing to really nurture them, or to provide the physical care, feeding, bathing, or the 'spoiling' (*manja*) that old people, like children, are thought to require. 'Daughters-in-law? Well, some are lucky, but that is rare. They can take care of us for a few days, but if we are ill for a long time, they must begin to complain.'

Daughters are thought to be a more reliable source of support in old age than sons because of the emotional bond between mother and daughter. But unless they are working, daughters are dependent on their husbands financially and may not be able to help. Dependence on a son-in-law seems to be particularly untenable for fathers, who state that if they have to depend on anyone, it is better to depend on their own son since 'he is head of the household, he's the one who earns the money, so his wife can't complain'. Both mothers and fathers express a clear preference to live on their own because 'it is troublesome to live with children'. A survey carried out in Singapore shows this desire for independence, and also the difference in the preferences of men and women (Table 4.1).

Since children are expected to be unreliable, parents pursue various strategies to ensure security in old age. A most important one is personal savings, and this is reflected in a finding of the same survey (Table 4.2).

Savings can be used in two ways. They can be used to pay for rent and food, or they can be used as a 'hold' over children. The latter strategy is more often used by women, especially those whose relationships with their children are not good. Money, gold, or land can be given to the person who cares for the old, and the promise of financial reward helps

TABLE 4.1
Malay Plan of Living Arrangements in Old Age
(per cent)

	Female Respondents	Male Respondents
With a son	14.6	23.4
With a daughter	23.6	10.9
Alone with spouse	59.7	62.5
Other arrrangement	2.1	3.2
	100.0	100.0

Source: Adapted from Chen *et al.*, 1982: 35.

TABLE 4.2
Malay Perceived Sources of Important Financial Support in Old Age
(per cent)

	Female Respondents	Male Respondents
Personal savings	87.4	77.8
CPF/Pension	6.3	19.0
Children	6.3	1.6
Other Sources	0.0	1.6
	100.0	100.0

Source: Adapted from Chen et al., 1982: 37.

to ensure that such care is forthcoming. Owning a good house is a similar form of security, since one of the children will probably choose to live in the house and the old people living there will be given food by the child with whom they have established co-residence. Most of the present older generation of Singapore Malays own neither houses, land, nor substantial savings, having raised large families on low incomes and received little return from their children. They must therefore rely on other strategies, one of which is directly related to the size of their family.

If no *one* child can be entirely relied upon, because he or she may be unwilling or unable to support the parent, then a strategy to increase old-age security is for the parent to multiply the number of children, and hence the number of possibilities of support. This reasoning is explicitly stated:

We like to have a lot of children, because when we are old, we don't know which one will want to look after us. If we have just one or two, maybe the in-law will be no good, then where will we stay?

If we have many children, we can stay in one place, then if we fight, we can move into another. Many old people do that. They circulate (*pusing*). It always turns out like that. People can't get along.

Out of ten children, maybe two or three will remember their mother and look after her when she is old.

This is an interesting variation on Caldwell's argument that it is the flow of wealth from child to parent that encourages high fertility. In this case, it is not the flow of material wealth upwards but old-age care that is the parents' primary and overwhelming concern. The care provided by children for their parents in old age is of incalculable economic and emotional value to the parents, and it sways parental decisions in favour of high fertility despite the downward flow of wealth from parent to child throughout the other phases of the life cycle. Moreover, it is not the guarantee of return from children but precisely the lack of guarantees that makes this multiplication of filial bonds rational. High fertility is not the only means of multiplying filial bonds, since it is not biological parenthood but social parenthood, or the establishment of emotional bonds through the provision of nurture to a child, that imposes an obligation to care for the old. Adoption or the raising of grandchildren is a popular and sometimes explicit strategy to increase the number of possible sources of old-age support. Old ladies usually receive regular contributions of cash from grandchildren or informally adopted children they have raised in whole or in part, and from the small gifts of money from a large number of 'children' they may amass significant savings in their old age.

When the economic aspect of these relationships is made clear, it is usually a sign of disappointment and frustration, as in the widely used expression, 'one mother can look after ten children, but ten children can't look after one mother'. However, the cultivation of emotional bonds with children is not only an economic strategy, since kinship sentiment is valued in itself. The old may be distressed by their lack of funds, but they are even more upset by the lack of funds as evidence of a breakdown of kinship sentiment. Their greatest pride and joy is to have children who visit them frequently and always see them when they are ill, and material gifts are viewed as tokens of the children's love and concern. One old man, very content with his grandparenthood, indicated that he was not concerned with amassing money or exercising authority but with being wanted and loved by all his children and grandchildren. He described his situation:

I stay with my sister's child; then my child, this one here, says, 'You can't stay there, you must come and stay with me,' but the one who lives here doesn't want to let me go, so I go there for a few days, here for a few days, then my other child says, 'If you want to come over here, just phone and I'll come and collect you with my car,' so I go there, too. They all tell me not to work, they want to give me money; they say, 'Just stay at home or go where you like and come back to eat; there is no problem.'

In fact, all this man's 'children' are adopted, which makes his success more remarkable. Love, care, and material resources which he had voluntarily bestowed on children are being channelled back to him, and this is a source of personal pride, and proof to all onlookers that he must be a really good and lucky man.

The younger generation of Malays in Singapore are clearly basing their fertility decisions on a different set of factors from their parents, since the fertility rate has dropped so dramatically in just 20 years (see Table 7.7). All the younger-generation men and women who have worked will have substantial savings for their old age under the compulsory national saving and pension programme (CPF), and most young couples are purchasing their own homes under government schemes. They do not foresee any need to depend on their children financially in old age, and while they value the love and care of their children just as highly, they appear to expect to achieve a close bond by concentrating emotional intensity on a small number of children. They are also careful in the religious training of their children, and note that Islam stresses the duty of children and all Muslims to care for the old and infirm.

It was suggested earlier that the emphasis on emotional bonds to children among the present older generation of Malays in Singapore is related to their lack of property. Those who have no property to support themselves, or to use as a hold over their children, have only emotional bonds to rely on, however unsatisfactory or insecure. Kinship sentiment is of great practical importance to the poorer wage-earners in the city. A comparison with the situation in the rural Malay hinterland will show that there, too, security is obtained through the ownership of property, and the rural poor cannot rely on their children.

It is a common practice in Java and Malaysia for the old to postpone transfer of the title to their land until after death. Only the usufruct rights are transferred to children on marriage since, according to a writer on rural North Peninsular Malaysia, 'apparently parents are not enthusiastic about giving their land away to their children because they are afraid that once they do so, their children will not take care of them' (Diana Wong, 1983: 277). Another writer on rural Malaysia states that 'since kinship reciprocity is voluntary even though care of one's parents is required in moral law, there is no guarantee that a father who provides little for his son save good will will receive any security in his old age' (Banks, 1983: 157).[1]

There are several ways in which property is used to guarantee old-age security. Jay (1969: 77) found that in Java, parents either withhold some property rights as a hold over the younger generation, or hand over most of their capital goods, retaining a piece of land specifically for their own support. This land could be worked by the parents themselves, or by the children or someone else on a share-cropping basis. A similar arrangement is made in the Malaysian fishing economy for the transfer of boats and nets (Bailey, 1983: 183). The understanding may be that the child who share-crops the parents' land will gain title to it on their death, but in the meantime parental support is ensured on a quasi-contractual basis and not left to the vagaries of kinship sentiment. Parents can always disinherit recalcitrant children by giving the property to others. While rights to much of the parents' property are transferred to their children before their death, it is common for parents to bequeath their residual assets to the person who cares for them in old age, who may be a grandchild or adopted child rather than their own offspring.

Diana Wong (1983) has described the process by which rural Malay parents with assets, land, and housing, are able to keep their married children close to them and ensure old-age support, while parents with no assets systematically 'lose' their own children as they go to live with or near their in-laws. If neither side can provide the young couple with rural capital, they migrate out of the area entirely. The decision on whether a young Malay couple will be oriented towards the side of the bride or the groom has generally been described in the ethnographic literature as a 'choice' and discussed from the viewpoint of the couple. Diana Wong points out the implications of this decision for the parents, and finds that the 'loss' of children upon marriage to the side of the more wealthy spouse routinely leads to the emergence of denuded families. In the village she studied, there were thirty elderly people aged 60 and above. Of these, half lived in a stem family structure with one of their children, while half were still struggling to make ends meet, often with a grandchild or spouse, or an unmarried or divorced child. Ten of these denuded households were female-headed, and all the women had married and had children living in the village or nearby (Diana Wong, 1983: 273). The old people in the village studied, obtained food by their own efforts, supplemented by the charity of other villagers and some contributions from their own children. In contrast, parents who had been able to provide their children with usufruct or share-cropping rights had provisions for their own support and a much better chance of establishing co-residence with a married child, which gave them the right to consume cooked food in the child's household.

Since land is accumulated by the joint efforts of husband and wife in the rural economy, divorcees are particularly vulnerable to the 'loss' of their children on marriage. According to Diana Wong, a divorcee with little land can return to live with her own parents but she is unlikely to be able to accumulate sufficient land and other assets to provide her children with access to agricultural production. If she remarries, her children may be left with the grandparents but in 'either case, upon the children's marriage, the new couple will be integrated into the parental unit of the in-law, and feel little obligation to provide for the old-age needs of the mother' (Diana Wong, 1983: 280).

Diana Wong's research on the position of the elderly in Malaysia suggests that the insecurity of the old in Singapore is not the result of a breakdown of a rural kinship system in which the old would automatically have been cared for by virtue of their status, authority, or participation in a corporate family economy. Rural villagers frequently make arrangements to feed and to bury the elderly poor, indicating that the neglect shown by children is not unexpected, though neither is it fully institutionalized nor condoned. It is still felt that ideally, the old should be taken care of by their children. Children are criticized when this care is not given, but at the same time children who do care for their elderly parents are the subject of positive commentary, because a close relationship sustained to old age is an achievement on the part of both parent and child, and not something taken for granted.

Housing conditions in Singapore prevent the elderly from living alone as they often do in rural areas, and while this may force them into socially uncomfortable living arrangements with children and in-laws, it does secure their daily consumption rights. Urban children are probably more aware of a *duty* towards their parents than some rural Malays, since they are more exposed to Islamic proselytizing. In both rural and urban contexts, the economic assets of the old help to draw their children close to them, and it is sometimes acknowledged that an element of self-interest, especially the expectation of inheritance, underlies the children's attentions. Giving away their wealth generously is also one of the ways in which the old create kinship sentiment. Old people who retain all their riches, having alienated their children, are considered unnaturally mean and unwise, as they will probably die alone and unloved. A successful parent lives out old age cosseted by devoted children, and though wealth is one way to secure their attention, it is kinship sentiment itself that is really valued.

The Personal Life Span and the Afterlife

Singapore Malays look forward to their old age not as a time of material plenty, but as a time when they are released from the responsibility of raising, educating, and marrying off their children, and they can focus on prayer and the afterlife. The relationship between parents and children continues after the parents' death, and research in Singapore suggests that the flow of material goods and loving kindness remains the principal link between the generations, although the forms through which these are expressed are of course different. The key benefit that a parent could potentially obtain from a child after the parent's death, is assistance in finding happiness and tranquillity in the afterlife. The child could obtain material goods (inheritance) and parental blessings. The extent and nature of these transfers are examined here.

In Islam, salvation is basically individual. A pious Muslim can secure his or her salvation by adhering to religious law, and carrying out the five daily prayers and the annual fast. A Muslim is not enjoined to perform the pilgrimage if he or she does not have the means. Beyond personal conduct, the major obligation of a Muslim parent is to see to the religious education of children until they reach maturity, whereupon boys take on responsibility for their own conduct and a girl's burden is shouldered by her husband. Not all Muslims in Singapore actually carry out the five daily prayers, though most *do* fast. Their diligence in prayer increases with age, and at the later stage in their lives many become active in groups formed for additional prayer. This is intended to add to their credit of personal religious grace (*pahala*), or decrease the deficit from earlier years.

There are two further potential sources of religious grace. One is religious works for the community, such as becoming a religious teacher or donating money to mosques, orphans, and various specified religious causes. The other is the prayers of children for the salvation of their parents after death. This is a potential transfer from children to parents that could make children incalculably valuable to their parents.

Children's contribution to their parents' *pahala* is subject to a range of interpretations. Reformist Muslims consider that a person's relation to this world, and to his children, is completely severed at the moment of his death, and there are no additions to, or reductions from, his accumulation of *pahala* after that point. They assert that the proper emphasis is on the religious education of children, since it is fulfilling one's duty to make one's children pious, not their prayers for the dead as such, that will be rewarded in heaven. Malays are Sunni Muslims and most orthodox exponents of this school consider that it *is* possible to intercede for the dead. Allah will accept prayers for the dead, though the granting of salvation is only at His will. Some interpret the word 'children' to refer only to biological children or children a person has raised; others interpret it to mean any Muslim individuals who are willing to join in prayer for the deceased. In the latter case, the good relationships a person has established with friends, neighbours, and kin during his lifetime will benefit him after his death, as these people will be willing to pray for his soul.

The most common Malay interpretation is that the more people who gather to pray for the dead, or who give donations that can be used to finance prayers over the years, the greater quantitatively the *pahala* of the deceased, and the lighter the sentence in purgatory. In either case, it is the children and spouse of the deceased who organize the prayer. Many believe there is an obligation on the spouse, and each child with a separate household, to host the whole series of prayers held 7 days, 40 days, and 100 days after the death and annually thereafter, to the extent that they can afford to do so. In Singapore and elsewhere, prayers are held for adoptive social parents in addition to biological parents (see Jay, 1969: 179).

The relationship between children and their parents after death is viewed as a continuation of the relationship established before death, and it is subject to similar strains and ambivalence. Parents do not stress that prayers for the soul are obligatory, since forced prayers are not considered effective. Prayer is valid only if it is voluntary, at the child's will. Insecurity is expressed over whether the child will in fact remember the deceased person, and continue to feel an emotional bond, just as there is insecurity over the relationship with children in old age. Children feel an obligation to perform these prayers, which they relate to the nurture received when young (hence the inclusion of adoptive parents). A few describe the obligation explicitly as a responsibility (*tanggungjawab*), but it is more frequently seen as help (*bantu*, *tolong*), as an act of generosity stemming from pity (*kasihan*), or even as a present (*hadiah*). Prayers for deceased parents are seen as voluntary counter-gifts, related to the parental gift of nurture but not reducible to clearly defined rights and obligations.

We feel sorry for the dead person (*kasihan*). Here we pray, and he receives the blessings over there. If we don't carry out these prayers, all the other deceased will receive blessings and our own father will be left out. He would feel offended, and we would feel ashamed. It is like being sick in hospital when everyone else has visitors and no one comes to see our father; we feel sorry for him (*kasihan*). So we make him this gift (*hadiah*).

The emotional bond developed with children which will make them remember and feel concerned and sympathetic towards the old, is the only means to secure these prayers. Like the relationship developed during life, the relationship after death is tentative and conditional, and there is a view that here, too, the more children, the greater the chances of receiving prayers from at least some of them. Religious education of children is expected to increase their sense of responsibility as well as the effectiveness of their prayers, though the provision of this education is a parental duty in itself and is not primarily seen in terms of direct self-interest. Example is also thought to be effective. If children witness their parents taking care of their elderly grandparents in their last years and after death, they will come to understand what is expected of them, though it cannot be explicitly commanded. This is the principle Djamour (1959: 144) describes as 'indirect generation reciprocity'. She noted that parents see themselves as giving unstintingly to their children, as they were themselves given to by their parents. In the course of this research, frequent comments on indirect generation reciprocity were encountered, but related most often to the care of seniors: if we take care of our elderly, maybe our children will see our example, and later take care of us.

Many Malays believe that parents can intervene more directly to secure the prayers of their children. This is the opposite pole from the reformist Muslim view that there is no further relationship between parents and children after death. It is usually considered to be pre-Islamic or non-Islamic, associated with *adat* in Malaysia and with the *abangan* tradition in Java. The belief is that the souls of dead ancestors are hungry and lonely, and so return to the house of their descendants to taste the food at ritual feasts. If they feel neglected, they appear in dreams and remind the children of the emotional bonds and outstanding debts that continue to bind them to their parents: 'I was not feeling peaceful. I couldn't sleep properly. I kept dreaming. It was as if something was bothering me, making me think and remember, so I held a feast for my father; it's okay now, there's nothing there. Our parents know how to make us remember them.'

In Singapore, the prayers for the dead, *Kenduri Arwah* for personal ancestors and kin, and *Kenduri Ruahan* held annually for all Muslims, are interpreted in these different ways. For some the purpose is prayer, and the prayer benefits the dead by adding to his or her stock of grace. The food is a token of hospitality and gratitude to those who attended, and is also seen as charity which is itself a source of *pahala* to the donor and to the person in whose name it is given. For others the food itself, or its essence, is consumed by the dead, and the purpose of the prayer is to ask personal forgiveness from the dead for misdeeds and to receive their blessings (see also Jay, 1969: 111). Reformist Muslims see the prayers in yet another light, as an occasion to calm and comfort the bereaved by prayer and the presence of concerned kin and neighbours, which has no effect at all on the deceased but is beneficial to the Muslim community on earth. These purposes are not clearly separated and various interpretations are simultaneously present at one ritual and sometimes in the same person.

The continuing relationship between children and their parents after death is closely linked to the transfer of property between the generations. Here again, a range of interpretations exist. Under Muslim law, the descendants, spouse, and other kin of a deceased Muslim have rights to specific shares of his or her property. Intestate property is divided on this basis by the Syariah Court, and according to the current law in force for Muslims, it is not possible to use a will to bequeath property to individuals other than those prescribed by Muslim law or to vary the prescribed shares (Administration of Muslim Law Act, 1970: 39). This makes inheritance an absolute right, not a personal gift, though the individual is permitted to make gifts as he or she chooses before death. The Syariah Court recognizes an alternative method of distributing property, which is by the agreement of all the legitimate heirs. This is the most common practice, so that only disputed cases are finally divided according to the prescribed shares. In dividing the property, the heirs feel obliged to abide by the wishes and intentions of the deceased. Some name a favoured heir, but it is said that parents do not like to see their children quarrel so they avoid disputes by making an equitable distribution, or by actually giving away their property before death to settle the matter.

Whether the living receive property before or after the donor's death, by Islamic shares or some other arrangement, many Malays do not regard this transfer as a right but as a gift from the original owner to a descendant. Since it is a gift, those who receive it do not feel they wholly own it, as they would in a cash transaction or market exchange. The gift carries a trace of the giver, and sets up a permanent relationship of debt towards him or her which cannot be severed. This debt is thought to incur the obligation to make return gifts in the form of prayers for the dead. This is a personalized conception of property transmission which is related to the individual rights to property and labour power that a person has during his lifetime. Inheritance is seen as a gift from an individual, not as a corporate or even individual right. For those who believe in the continuation of a personal relationship with the deceased, it is the flow of material goods, gifts and their traces, that is the vehicle for constructing and sustaining a relationship with another individual after death, just as it is during life.

Descendants living on land cleared or originally purchased by an ancestor say they 'hitch a ride' on his or her wealth (*tumpang hartanya*), which incurs an obligation to remember him or her by offering annual prayers. If rent is received from the land or if the land is sold, the cash gained or at least a portion of it should be used for the afterlife of the deceased, because 'it is his money, not ours'. The money can be used to hold prayers, to donate to mosques, or to perform a posthumous pilgrimage for the deceased. Only the remainder can be used by the descendants, though the portion of the funds allocated to religious purposes is a matter for their conscience. Some believe that a curse will be placed on those who use too much of their ancestor's wealth for themselves, and this can take the form of illness, death, or poverty:

If we live on their land, we must hold a prayer for them every year; so, too, if we depend on the proceeds of their land, like the rent. It is as if the old people are feeding us, and we must remember them. If we don't, we will be cursed. Like those landowners over there. They did not remember, so now they have really fallen low and have put all their gold in pawn.

We pray for him every year, because he gave us this house to live in. If we don't hold the prayers, it would be like we are ungrateful to him.

The ability to curse the living provides parents with stronger mystical sanctions to reinforce their hold over their children's emotions and actions after death than they had while alive, as direct retribution will fall on children who fail to meet their obligations. Property provides the basis of security for the deceased, just as it does for the elderly. If property is given to children, they remain obligated to the deceased, and are reminded of him or her and of the outstanding debt every time they see or use the property. 'It is like this. If we leave land, then when our child goes along to look at the land with his own child, the child is bound to ask, "Whose garden is this?", since children are always curious. Then the father will tell him, "This belongs to grandfather so and so", and the child will know about us.' Property thus serves as a physical reminder of the deceased which keeps the emotional bond alive. It is the emotional bond which is the source of the gratitude and sympathy for the deceased, which makes descendants *want* to help him or her with prayers. Though the strength and continuity of emotional bonds cannot be foreseen, commanded, or guaranteed, property can help to secure it.

There is no doubt that the beliefs described above remain widespread among Singapore Malays. During the last ten years, all the Malay-owned kampong land in Singapore has been taken over for redevelopment, and in all the cases where information was obtained, some of the money from the sale of the land had been put aside for prayers for the souls of the original owners or clearers of the land, as well as the grandparents and parents through whom the present owners came to have legal rights to the land. Other Malay property is treated in a similar way. Many believe that money left by the deceased should not be used by the descendants, since it does not belong to them. It should be used for the afterlife of the deceased owner. So, too, with gold accumulated by women if any of it is sold, though descendants can make use of it in the meantime (*tumpang*). However, it is always stressed that only descendants who remember the deceased with concern, affection, and gratitude for his gifts which give them wealth or income, will treat his property and his soul in this way. The return to the deceased is conditional and not guaranteed. It depends on personal relationships and on sentiment.

The importance individuals ascribe to the prayers of children after their death varies widely. While all parents are concerned about sustaining a relationship with their children during old age, not all are very concerned about their afterlife. Those who are most knowledgeable about canonical Islam rely least on their children or other people, and prepare themselves for the afterlife by their personal conduct. Others, who hold some of the beliefs described here about a continuing relationship between parents and children after death, are often not deeply concerned about punishment in the afterlife. As Djamour (1959: 146) noted, the 'majority of them appear to think that if they have committed no major sin, such as murder, and if they are given a proper Muslim burial, their souls will rest in peace'.

For some Malays who are concerned about the suffering of their souls, the prayers of their children are of inestimable value. For others, less concerned about the suffering of their souls, it remains to clarify *why* they are so diligent in holding ritual feasts and prayers for the dead. The preceding analysis of the relationship between parents and children while the parents are alive and after their death, suggests the interpretation that kinship sentiment is really the primary issue. It is continuity of the emotional bond between parents and children that is demonstrated and reinforced by the exchanges of food, prayers, and blessings that take place at these feasts. The attitudes towards the dead and towards their property, described above, rest upon the conviction that parents want to feel loved by their children, even after death. It is this love that children continue to bestow on their parents at the final stage in the family cycle, and it is a gift that is highly valued. The *kenduri* is an occasion on which the children and spouse of the deceased demonstrate publicly that they remember the deceased with love and gratitude. This shows that both they and the deceased are good and worthy people who have accomplished and enjoyed a relationship built out of kinship sentiment and generosity. Both in the rural Malay setting and in Singapore, this accomplishment is the foremost source of self- and social-esteem.

Kinship Sentiment and the Value of Children

This chapter has described the cultural basis on which parents and children bring together the variety of cash, goods, and services that sustain not only the current producers, but also the young and the old through the various stages of the life cycle. It described the interplay of ideas of obligation, self-interested calculation, and kinship sentiment, which are ideas that draw on the Malay cultural heritage although they are not simply reproduced in unchanging form. In a way that is at first sight paradoxical, the economic conditions in Singapore, in particular the commodification of housework and the relative affluence of the younger generation, have helped to reinforce kinship sentiment, which is built out of voluntary exchanges between individuals. The prominence of kinship sentiment in the relationship between parents and children is not false consciousness nor merely a deceptive idiom. It is the result of the creative application of cultural knowledge to new situations, as Singapore Malays have tried to organize their lives in ways which seem to them gratifying and satisfactory.

The pattern described in Singapore does not stem inevitably from either the Malay or Javanese cultural heritage or current material conditions, since it is possible to picture quite different arrangements. For example, parents who have become aware of the monetary costs of raising children in the highly commercialized economy, could stress their individual right to recoup their loss. They could assert that their children only gain access to the fruits of their individual labour once they have discharged their monetary debts to their parents. This would be the logical conclusion of individualism, but it would have a cost, in that debts fully discharged could sever social relations entirely, leaving the old isolated. In the individualistic kinship pattern in Britain historically, according to Macfarlane (1986), parents exercised the gift form more thoroughly, in that they regarded children as a total luxury and expected no return from them at any stage or in any form. They did not want crowds of mourners at their funerals nor prayers after death, and they expected to be cared for in their old age by servants whom they could pay from their own savings in a relationship of commodity exchange. It seems that Malays value kinship sentiment more highly, but also place more strains on relations with kin by the mix of calculation and generosity in their various transactions.

It may be appropriate to end this chapter on the relations between parents and children, with a comment on Caldwell's theory of intergenerational wealth-flows (Caldwell, 1976). He asserts that there are two possible regimes, one in which wealth flows from child to parent and high fertility is of economic benefit to the parent, and one in which wealth flows from parent to child and parents have a direct interest in restrained fertility. He considers the latter system to result from the adoption of the Western nuclear family system which is being transmitted along with its concomitant values by diffusion from the West to all other parts of the world.

The Malay and Javanese rural ethnographies discussed here show, first of all, that the Malay world is a cultural region which has been characterized by a flow of material wealth from parents to children in its rural history, long before the impact of the West. Even more interesting, and contrary to Caldwell's predictions, it seems to be the urban wage economy in Singapore which has begun to promote a flow of material wealth in the opposite direction, that is, back towards the parents, as Malay mothers in particular seek to gain a share of their children's new-found affluence.

There are two difficulties with Caldwell's analysis. The wealth-flows upon which he focuses, tend to be viewed in the narrow sense of direct transfers of material wealth, and he notes, but puts aside, the question of the 'non economic' values of children. The analysis presented here explores the theme of the costs and value of children, but does so by taking into account intangible values such as the emotional bond to children in old age, or even after death, and practical questions such as the opportunity costs of the mother's labour time, to herself and to her family, the availability of older sibling baby-sitters, and the importance given to the accumulation of material wealth versus other goals. Value cannot be reduced to material wealth but must itself be subjected to cultural and historical analysis.

The more general weakness of Caldwell's argument lies in the tendency to reify culture, and treat it as a 'thing' which has effectiveness in and of itself and which can be transmitted by diffusion. He treats the 'logic of sentiment' as the motor of historical change (Thadani, 1978).

His work does not pay sufficient attention to analysis of the way cultures constantly change and develop in relation to their specific past and to new conditions. There are not, as Caldwell asserts, only two possible types of family system or a universal direction of change, but many variations. It is possible to shift from a system where the predominant wealth-flow is downwards towards children, to one where wealth begins to flow towards parents, and then change again. New cultural ideas and new economic systems such as those imported from the West do intrude on daily life, but as part of a continuing, unique, and specific history, not as final and definitive imprints on a passive or unstructured social and economic setting. It is the active and creative moment in which cultural ideas and economic conditions are woven into the practices of daily life, that has been explored in the course of this analysis of Malay householding.

1. See Macfarlane (1986) for an interesting account of a similar ambiguity and anxiety in the relations between parents and children in England historically. There, too, the rights and duties of parents and children were not clearly specified, though moralizing texts suggested that children did have an obligation. Parents were advised not to give away all their property too soon, leaving themselves as possibly unwelcome lodgers in their children's or other people's homes.

5 Householding Relationships and the Expansion of Economic Resources

THE way householding is organized, has a structuring effect on the economic position of the individual and the household in the national economy. This chapter examines some of the consequences of the householding pattern described earlier for the expansion of the material resources of the individual or the householding group. It considers the extent to which Malay householding relations provide the incentive, the encouragement, and the practical means for the expansion of the individual's, or the household's, economic resources.

The preceding discussion of the relationships between husband and wife, parent and child, has shown how the current household is sustained and the new generation raised to adulthood. Though there are ideas of duty and obligation, it was argued that much of the transference of resources, cash, and services between the parties is seen as voluntary and as a gift made at a direct personal loss. There is no concept of a joint household productive base, or even a joint dyad. The resources generally made available to other household members are only those necessary for current consumption, or for extra gifts which form the material and emotional currency sustaining the channels of exchange between individuals.

In this system, no two individuals have a shared economic interest in expanding their economic resources. Husband and wife may be prepared to undertake a joint economic venture but this tends to be on the basis of an explicitly negotiated partnership, shared profits, and restricted in extent and duration by the fear of divorce. Siblings do not inherit from a joint fund, since transfers to children are made at parental whim, so their economic incentive to contribute their individual efforts to expanding the resource base of their siblings is absent both before and after marriage. Likewise, if children contribute to their parents' accumulation, they have no definite right to expect any return since the gifts they make to their parents are voluntary and final, and parents make gifts to their children at their own discretion. Parents may be prepared to use their personal resources to establish their children with education, capital and other material assets, but the economic incentive they have for doing so is limited by the conditional nature of the parent-child bond.

These are the implications of the system of individualism described, but they do not wholly determine the ways that individuals choose to act. Choices made are affected by material opportunities and by other values that individuals hold such as religious values, the positive desire to see loved ones prosper, and pride in personal achievement. This chapter examines the extent to which individuals seek to accumulate material wealth, the reasons they have for doing so, and the difficulties stemming from within the household that impede their attempts. It focuses on three areas: the question of individual salvation and the dispersal of inheritance, parental investment in children, and the role of the family in business enterprise.

Individual Salvation and the Dispersal of Inheritance

The extent to which an individual seeks to accumulate wealth during his or her lifetime depends, in part, on the way that wealth is expected to be used after the individual's death. If the wealth is to be used for the individual's own salvation, the individual has a definite incentive to accumulate. There may also be an incentive to accumulate wealth in order to leave it to the next generation, but this depends on how the relationship between the generations is seen. If wealth is accumulated and passed on, the next generation has the advantage of beginning its own accumulation from a stronger resource base, promoting the expansion of economic resources across the generations.

In the Malay case, there is a range of views about the way salvation in the afterlife is attained, and the role that material wealth plays in securing salvation. On the positive side, material wealth can be donated to mosques and other religious causes, an act which is a source of grace to the donor. Wealth can also be used to finance feasts at which descendants and others pray for the soul of the dead, an act which many believe lightens the suffering of the deceased soul. In both these instances, the individual has an incentive to accumulate wealth, since it will be used primarily for his own benefit. Wealth gained by the efforts of an individual is used for that individual's salvation. Though it may pass through the hands of heirs who make donations and hold feasts on the deceased's behalf, a major part of the wealth of the deceased is dispersed to religious causes and does not become a part of the resource base of the next generation. Part of the wealth may be used by the descendants, though as described earlier, ultimate ownership and the benefit of the prayers of the descendants, are thought to belong by right to the individual who originally accumulated that wealth.

Muslim reformers criticize the Malay emphasis on using the resources of the deceased for feasts, as a source of Malay economic backwardness. They say feasts are ineffective in increasing the grace of the deceased, and merely waste resources that could become capital or could be used for the education of his children. While Islam does enjoin the wealthy to donate to religious causes, they should never do so at the expense of the welfare of their wives and children. In either the reformist view or the more common Malay view, the proportion of the wealth used for the deceased, as opposed to that used by the descendants, is a matter of conscience and judgement. In terms of the afterlife, the individual's best interest is probably served most securely by making donations to religious causes before death, since children cannot be relied upon to observe their parents' wishes or to remember their obligations to the dead and may instead use the money themselves. Many Malays and other Muslims donate a large part of their capital reserves to religious causes before their death. This means that some, if not all, of an individual's accumulated assets are dispersed at the end of the individual life span, and are not available for the use of the next generation.

Many Malays believe that wealth can play a negative role and impede the attainment of salvation, and that an excess of wealth will cause the individual to sin. Apart from providing for current consumption needs (though the standards considered comfortable or reasonable vary widely of course), individuals claim they have no use for more money: 'We Malays can't drink or gamble or go to night clubs, which is how the Chinese spend their money, so what is the point in chasing after more and more?' Implicit in this statement is the idea that wealth is primarily intended for current consumption during the individual's life span. It is not seen as an accumulating fund that will continue across the generations. This one-lifetime view of the utility of wealth is shown in the critical remark directed at a rich profligate–'he has so much money, he can't finish eating it'–and also in the phrase often used in the context of the refusal or neglect of economic opportunities, 'we only take a white cloth when we die, we can't take money with us'. The implication is that it is not worthwhile for the individual to accumulate more wealth than he or she can use in his or her *own* lifetime.

These views again are criticized by Muslim reformers, who say that riches accumulated by an individual and passed on to his descendants can be used to good purpose. They can create more wealth through business or the provision of education, which in turn increases the fund that can be used to donate to the poor and other religious causes. Excess riches are seen as a potential evil by Malays partly because they generally think of them in terms of consumption, not as business capital, for reasons to be discussed later. Muslim reformers are often involved in business, and therefore can generally see productive ways in which large amounts of money can be used.

For many Malays, the negative potentials of wealth extend across the generations. For those who believe in a continuing relationship with their descendants after death, the misuse that the inheritors may make of an individual's wealth can damage his or her status in the afterlife: 'If our children sin with our money, it is we who are at fault.' Children are also thought very likely to quarrel over money, especially if there is a large inheritance, and many believe this will disturb the souls of the dead just as it upsets parents to see their children fighting. The reformist view is that proper religious training of children will ensure that they do not misuse wealth and if the inheritance is distributed according to the just shares decreed by religious law, it will not be the cause of quarrels.

This set of sometimes conflicting views does not conclusively determine whether or not individuals will seek to expand their material resources during their lifetime. Wealth can be a source of old-age security and contribute to salvation, but wealth is not necessary for individual salvation, as the pious poor also go to heaven, and given the temptation to misuse wealth, an individual may feel better off without it. There is an obligation to provide sustenance to offspring while they are under the parents' care, but there is no obligation to make them rich, and no definite individual advantage to the parents in the afterlife is served by doing so. Parents may still want to see their children comfortably established for other reasons to be discussed.

As an example of the way capital tends to be dispersed at the end of the life cycle, this is the manner in which one Singapore Malay distributed the $50,000 she received under the government savings plan on her fifty-fifth birthday: she used $8,000 to perform her *haj* (pilgrimage to Mecca); she put aside $5,000 each for her two sons and one adopted son, though she made the arrangement secretly with a lawyer for fear the children might otherwise mistreat her in old age; she kept $12,000 to meet her own needs, and she donated $15,000 to religious causes. An orientation towards personal salvation and ambivalence towards children occasioned the dispersal of funds which, seen from a different cultural perspective, could have been used to increase current living standards or provide business or educational opportunities for the next generation.

Parental Investment in Children: Land and Education

The term 'investment' is used here because it was shown earlier that Malay parents do expect, or at least hope for, some form of return from the resources they bestow on their children. This is not a direct material return. Parents hope that the children they have raised to maturity and provided with the means of gaining a livelihood, will repay the parents with affection, gratitude, and care in old age and after their death.

In rural areas of Malaysia and Java, land is the major productive asset parents bestow on their children. In order to provide their children with agricultural capital, parents must either endeavour to accumulate additional land, or be prepared to accept a reduction in their own resource base as they give their children land they are currently farming. The parents' resources and diligence, and the number of children they have, affect the amount of land they can give to their children, though the transfer is ultimately determined by the parents' willingness to make the gift. Banks (1983: 157) describes this transfer of resources as the supreme gift or act of kinship, and the essence of the parental role. He found (1983: 175) that where land has become scarce and parents are unable to establish their children with agricultural capital, 'the impression is left that the father has not performed his role adequately or well'. Parents bear part of the cost of this apparent failure, as they are unable to keep their children close by them and to secure support in old age, as noted earlier.

The resources relevant to the economy of Malay households in Singapore are rather different from the rural economy described by Banks, but in Singapore too the parental role in establishing the livelihood of the next generation has been undermined. This affects the extent to which parents are able or willing to attempt to maintain or improve the resource base and living standards of the next generation. Malays in Singapore are seldom involved in entrepreneurship. While in the rural economy any cash or capital goods parents give their children become part of the children's productive and potentially expanding resources, in Singapore cash, shares, gold, housing, or other forms of wealth appear to have limited uses, as wealth is seen primarily in terms of consumption. Some parents have no material assets to give their children. Those who do have assets, have often provided sufficient education for their children, enabling the younger generation to obtain good jobs and become relatively affluent, and hence less interested in the parents' wealth: 'The children don't want our money. They say we worked for hundreds, they work for thousands; they say "Take your money, sell your house, go on the *haj*; we don't want it." Even our gold, they say, is old fashioned and they prefer to buy their own. They want nothing from us.'

Performing the *haj*, in some cases numerous times, has become a way for old people to use up the resources that their children appear not to need. It is a form of consumption which gives guaranteed benefits to the old themselves in the afterlife: 'If we were in a rural village, we could build a house, or buy a bit of land and let the children or grandchildren live there. Here our children say they have already bought their own houses with their CPF, and they don't want our money. So we go on the *haj*. At least we get some blessing from God for our afterlife.'

The relative affluence of the young, and the non-entrepreneurial view taken of material assets, have undermined the role of parents in supplying or augmenting the resource base of their children. This seems to have resulted in a greater economic disjunction between the generations than is found in the rural Malay world where parents' economic assets and their willingness to transfer some of these to their children are crucial to the establishment of their children's means of livelihood.

Education is the only form of investment made by Malay parents in Singapore that could potentially bring the benefits of assured and possibly improved livelihood to their children, and the rewards to themselves of successfully accomplished parenthood and secure old age. But there are some crucial differences between rural land and education as forms of investment. First of all, education, unlike land, gives the parents no security. It was noted earlier that parents are reluctant to hand over land titles and other assets to their children before death because they fear that children may not take care of them in old age. The gift of education is made earlier in the life cycle, and provides no means of holding children to their obligation to support the old. Insecurity is expressed over whether children would remember their parents no matter *what* sacrifices were made for their education: 'We put out the capital to give them education, and they may pass their exams, but still, out of ten children, only two or three will remember their parents.'

Parents note that it is often not the best-educated or best-paid child who remembers and cares for them in their old age. The common belief that individuals who have become better off than their kin, neighbours, or friends will become proud and ignore their former peers, extends also to children, so that education, from the parents' point of view, may even be counter-productive. Even before the children marry and leave home, there is no guarantee that any of the material benefits of education given to children will be channelled back to the parents, through contributions from the children's income. The one-sided benefits from education are explicitly acknowledged by parents, and the non-expectation of return enhances the parental sense of generosity: 'We give them education so that their future will be comfortable; they are the ones who profit from it, not us.'

Educational investment is ephemeral in another way. It does not guarantee any return, even to the child. The child may fail, then the parents' efforts are wasted and the child is not set up for a secure future. Educational success and failure, unlike land and other property, is not a direct outcome of the investment of capital and labour. Neither is success a parental achievement, since it is determined by the child's will and ability. Parents whose children have done well are considered lucky (*untung*), an expression that would not be used for rural parents who had diligently built up a starting capital in land for each child. The success is vicarious, as was noted by one mother who claimed she was just a 'passenger' in enjoying her daughter's achievements.

Parents sometimes explicitly liken education to inheritance and state that education is the only wealth that they are able to bestow on their children. More often, though, education is perceived as part of the normal nurture and raising of a child, and the routine expenses for school books, transport, tuition fees, and the like are met out of the current budget. Apart from meeting these expenses, parents often feel there is little more that an investment of money can do to improve their children's educational attainment. Unlike land, education appears to have a fixed investment limit. The result depends on the child and on an educational system over which the parents feel they exercise little control.

It is possible that the individualistic family system described here does actually affect the educational attainment of Malay children. Malay parents, expecting to find an individual character and will in their children, seem ready to accept the opinion of a child who declares that he or she is 'already big' and no longer wishes to study. This is especially so for boys, over whom parental authority is weak, as noted. Parents see no way to force an unwilling child to study. More importantly, Malay parents do not bring moral pressure to bear on their children by pointing out that their parents have sacrificed, spent money, and expect them to complete their education so that they can return money to them, expand the family economy, or bring glory to family and kin. Instead, parents stress that they have fulfilled their obligation if they were willing to provide education; the result depends on the children's will and ability, and the good fortune, if there is any, will be theirs.

Besides the general pattern of relations between parents and children described here, other factors intervene to determine the extent and intensity of parental investment in children's education. All Malay parents in Singapore in the 1980s pay greater attention to their children's education than parents in the past, because the reduced family size, reduced divorce rate and reduced incidence of child transfers have intensified the interaction between the parent and each individual child. There are also class factors which affect all Singapore families. Differences in income between Malay households observed in Singapore did not affect the extent or nature of the parents' material expectations from their children, but middle- or higher-income families who have recently achieved social and economic mobility do take a more intense interest in their children's education, mainly for fear of embarrassment should their children experience downward mobility. Well-behaved and studious children reflect the class status of the family as a whole (see Chapter 8). Lower-income parents may invest in education in the *hope* that their children will succeed and later care for them in old age, but they do not risk embarrassment or loss of status if the child fails and they are consequently less preoccupied with education than their higher-income counterparts. Lower-income Malay parents do not expect or even hope to attain economic mobility through their children, since the children's future is expected to be separate from theirs. In view of the limited prospects of obtaining any material return from investment in children, parents of all socio-economic groups feel strongly that, in providing education, they are acting out of motives of generosity and love, and the positive desire to see their children more comfortable and financially secure than the parents had been.

There is a third difference between land and education as productive assets bestowed on children, that should be briefly noted here. It concerns the relationship between siblings. Rural parents pass on land and other assets to each of their children as they marry. The remainder is transferred in the parents' old age or after their death. Most Malay ethnographers note that, though the parents have the right to give to whom they choose, ideally all the siblings should receive approximately equal shares. It is much more difficult for parents to distribute the benefits of education equally among their children, and this raises the potential for sibling rivalry. It is rare for Malay parents to choose to withdraw children selectively from school in order to give other siblings the chance to study where resources are too limited for everyone. As an explicit parental strategy, this would provoke accusations of favouritism, as each sibling has a separate future and has nothing at all to gain from the superior education another has received. The fact remains, though, that pressure of resources or failure in studies will cause some children to withdraw from school earlier than others. Children living in the same household are in direct competition for educational funds, as this parental 'gift' is made while children co-reside, unlike rural land which is not transferred until the point where children are leaving the parental home on marriage.

The potential for sibling rivalry is made greater by the fact that it is often working siblings who are the source of the cash which parents use for the education of their juniors. Though the transfer is made via parental hands, the siblings are still aware of the flow of resources from one to another. Some Malay siblings in Singapore are willing to contribute to each other's education, to take responsibility for their siblings' needs, or to help out, but many others are not. Where siblings do not help, this is not regarded by the struggling parents or by the sibling whose education has suffered from lack of funds as particularly surprising, though it may be frustrating: 'They have their own expenses, they want to use their own money.' Financial assistance for education is even more scarce after the sibling has married, and this is both foreseen and accepted, since the needs of spouse and children take precedence over those of siblings.

To avoid sibling jealousy, some students in higher education try to finance themselves by working during vacations even though their parents or siblings may be able to provide for them. Students fear that the help siblings give will later become a cause of tension as 'old matters will be raised' (*bangkit*), and the sibling will say, 'Look, I paid for your education, because of me you are well off now.' It is not that the sibling would or could directly claim a share of the proceeds of his or her investment, but rather that the sibling's gift would engender a lasting sense of personal debt and obligation which is uncomfortable between siblings, though acceptable between parent and child. The common Malay idea of the burden of debts of kindness (*hutang budi*) applies to the sibling relationship. A 70-year-old man who had helped to finance his younger sibling's education forty years earlier described this sense of lasting obligation: 'My brother wanted to stop school, because my father had retired. Luckily, he telephoned me, so I said no, you continue, and I paid for him. If we do good to people, they will remember us, whether they are relatives or other people. Till now, if I meet my brother in the street, he gives me $50; I say there's no need to, I've got money.'

Described above are some of the limitations on the parents' ability and willingness to provide their children with a fund of capital goods or education that will enable the new generation to live at a better standard than the old. The relationship between parents and children is organized to produce a minimum accumulation in parental hands, and this reduces the funds that can be bestowed on some or all of the children. If parents accumulate assets to give to their children, they do so by their own efforts and primarily out of their desire to augment the gift that they can bestow. Siblings have no economic interest in helping each other to accumulate assets or education, either directly or via the parents, though siblings, too, may choose to be generous.

The Malay Household and Business Enterprise

The preceding analysis indicates some limitations on the incentives Malay individuals have to accumulate wealth and assets. Neither salvation in the afterlife nor relations with children, are determined primarily by the accumulation and transfer of wealth, though wealth is not altogether irrelevant. The relations between parents and children, husbands, wives, and siblings, are arranged to produce the minimum pooling of funds between individuals. Examined here are the implications of this family system for Malay participation in business enterprise.

Business is a field of economic endeavour in which the motivation to accumulate and expand is central. A business enterprise can provide the entrepreneur with a regular income to meet current consumption needs. Business income is then treated as a salary, although it may be more lucrative, and it has the added advantage to the entrepreneur that he or she is self-employed and not subject to control and supervision. These are the reasons Malays most often give for entering, or planning to enter, business. In order to operate a business as an *expanding* enterprise, further motivation is required. This could be the desire to be very rich, or the desire to establish an estate that could be passed on to descendants who may carry on the entrepreneur's name, or a religious motivation. Only with such additional motivation will the entrepreneur find it rational to reinvest, to roll back funds, to restrict current consumption for future goals, or to take greater risks. All entrepreneurs do these things to some degree, but the extent to which they attempt to achieve maximum, perpetual, and limitless growth in their enterprise, varies with the strength of these motivations.

Weber (1930) argued that it was religious motivation which inspired the European entrepreneurs at the early stage of the development of capitalism in Europe. Clifford Geertz (1963: 49) considers this religious motivation or Protestant ethic to be present in Islam, and argues that Islam, too, enjoins the accumulation of wealth, the ascetic life-style, and the personal discipline that were found in Europe. Though it was noted earlier that many interpretations of Islam exist in Singapore, most Malays would agree that Islam favours entrepreneurship. Singapore Malay businessmen, just like the *Santri* observed by Clifford Geertz in Java, are often more religious than the average, and they claim that Islam provides a proper moral framework within which business can be carried out. Unlike the merchants described by Weber, however, Singapore Malays do not consider wealth to be essential to salvation nor is it proof of holiness or social or moral worth. Muslim businessmen are concerned to avoid infringing religious rules and they try to conduct their businesses in a steady, honest manner, taking few risks and gambles, avoiding debts and bankruptcy. Their strongest motivation to expand their businesses is pride in their personal achievement, and this is weighed against various disadvantages to expansion such as the loss of time to spend on prayer or with the family, and some social costs (see Chapter 9).

Neither the structure of relations between the generations, nor religious doctrines and perceptions of the afterlife, provide Singapore Malays with strong motivations for the accumulation of boundless wealth through business enterprise. Furthermore, the structure of householding relations inhibits the ability of the would-be entrepreneur to initiate and build up a business enterprise. It was shown earlier that Malay individuals are not able to draw automatically on the labour or financial resources of household members or other kin, even their own spouse, child, parent, or sibling. Any contribution made by family members has to be negotiated. In Singapore, Malay family members who contribute labour or capital to an individual's enterprise do so either on the basis of a gift, to 'help out' in the spirit of kinship, or on a contractual basis for which they are recompensed.

Family members help an entrepreneur with the gift of their labour in cases of temporary or short-term crisis, such as preparing extra food for sale before a national holiday. In return, the entrepreneur will usually show gratitude for their generosity by means of a token payment or counter-gift. If the labour requirement is longer-term, wages are paid. This is most common in a relationship between parent and child. The parent gives a wage in recognition of the child's desire for cash, and knows that without this attraction, the child will choose to work elsewhere where a wage is forthcoming. The freedom to allocate and enjoy his or her own labour power is considered to be the child's right. If the business is small and the wage is low, only children with low education who cannot command a high salary on the open market are attracted to a parent's business. A wage or fee is sometimes given between husband and wife, though it is often thinly disguised as cigarette- or pocket-money. It is rare for siblings to work for each other, especially after marriage, when it is expected that even a close, co-operative sibling relationship will be disrupted by the spouse, who is likely to complain that the hours are too long or the pay too low.

The employment of a spouse, child, or sibling in a Malay business is made difficult by a general reluctance to exert direct authority. Parents complain that their children are stubborn and do not obey orders. Few boys are happy under the constant, direct supervision of their fathers and, as noted, avoidance is preferred. There is no cultural stress on authority between siblings. Though younger siblings should show formal respect for their elders and listen to their advice, they are not under their command or control and strongly resent any attempts to exert such authority. These relationships do not provide a good basis for the running of a busy enterprise where commands often need to be abrupt and explicit.

Family members seldom invest capital in each other's businesses. They may lend cash to help out temporarily and this generosity is likely to be recognized by a token counter-gift when the sum is repaid. A contractual loan on which interest is charged is not allowed in Islam, and though an investment in a business enterprise which produces an income or share of profits is allowed, this practice is rare among Malays, since many believe Islamic law forbids this. The main disincentive to borrowing larger sums of capital from kin, is fear of a lasting personal obligation or debt. This syndrome is known as *bangkit* or 'bringing up old matters'. The idea was mentioned earlier in the case of aid between siblings in the provision of education. Even after the loan is repaid, the recipient feels obliged to recognize the lasting debt by continuing to give token gifts to the individuals who helped him start out, including his own parents: 'My aunt gave me $2,000 and my mother $3,000 when I needed extra capital to start this business. I have paid them back, and every year I give them some extra. I'm afraid of the day when they might turn and say "We helped you but you never gave us anything."'

The loan of money to the entrepreneur by the aunt and mother was a timely act, made out of generosity and concern. The personalized motivation, and the uniqueness of timing, made the loan into a gift which was unrepayable no matter how much the entrepreneur gave back at a later date. The resulting debt was felt to be onerous to the individual in his pursuit of his own material goals, and if the bank had been prepared to lend to him he would have preferred a formal contract that carried no lasting social obligation. He used the bonds of kinship sentiment to achieve his goal, but felt he lost some of his independence as a result, making the transaction a source of ambivalence and tension that could potentially disrupt the kinship bond itself. Paradoxically perhaps, while the idea of the gift in Malay culture is based on the concepts of individual integrity and choice, the obligations engendered by gifts reduce personal autonomy and tie the individual into social networks.

Though it is quite clear that individuals have no rights to the labour and reserves of their closest kin, Malays seldom overcome or resolve this limitation on pooling resources by the use of formal contracts that bring individuals together on the basis of a negotiated mutual agreement. The Malay marriage contract, especially in rural areas, could be regarded as a contractual business partnership and it is directly linked to the joint efforts of husband and wife in seeking to develop their agricultural assets. But it was argued earlier that the marriage relationship is reduced to this calculative, contractual, and self-interested element only when divorce is imminent. At other times, the couple are involved in exchanges which build and convey kinship sentiment, and the crucial feature of this sentiment is that calculations are not made explicit. To be involved in an openly calculative business contract with a spouse, child, or sibling seems to threaten the kinship basis of the whole relationship. This is especially so in the smallest businesses where there is no clear separation of the place of business from the household, and in some cases there is no separation of business and personal accounts.

In the Western world, where the kinship system has similar individualistic tendencies, it is common for a separation to be maintained between family and business accounts, and yet also for the family to be *incorporated* on a contractual, voluntary basis as a business enterprise. In the Malay case, the legal framework within which this could be done did not develop indigenously and the business and family *persona* are not separated. The explicit calculations essential to entrepreneurship are incompatible with the Malay idea of kinship sentiment. McKinley is correct in arguing that the absence of Malay participation in business on a family basis has to be seen in terms of the positive goals that individuals are trying to achieve. He criticizes the tendency to describe the Malay household or kinship system in relation to business by a series of negatives, such as the inability to co-operate, the absence of corporate descent groups which could sustain co-operation, ego-centrism, and so on (see Dewey, 1964: 237; Djamour, 1959: 46). McKinley (1975: 34, 35) counters that the Malay pattern is an:

> ...almost deliberate attempt to keep economic relations, as such, out of the direct affairs of people who are morally bound to each other. The situation is not one of marked individualism, so much as it is one of protecting certain very crucial moral bonds from disruptions often brought on by obvious material conflicts of interests.... [It is] a mechanism for maintaining a moral sphere of relationships, *as moral relationships*, and not a structure of incipient atomism.

Contrary to McKinley, it is argued here that economic relations *do* enter the affairs of people morally bound to each other. They do so on the basis of the daily and long-term exchanges of householding, and not only when families are involved in entrepreneurship. But the key point, which McKinley appreciates, is that economic relations *as such* or *obvious* material conflicts of interest are disguised, sometimes consciously, and at other times denied or negated, through acts of mutual generosity. This is the process that builds up those crucial moral bonds which do not exist without positive nurturing.

Some examples of Malay family involvement in business will illustrate this discussion. A typical small business is the making and peddling of cakes by women. It is very common for two or more women in one household, such as mother and daughter or daughter-in-law, to make and sell cakes separately. This separation occasions no comment. When asked why they do not join together, the answer is that they could work in partnership, but that would raise questions of whose cakes were tastier, took more labour or electricity, or sold better. Partnership is felt to be troublesome though there is a reluctance to specify further just what the source of trouble would be, especially in each other's presence. The father, husband, or child who peddles the cakes is paid a fee or earns commission on the cakes sold. After acknowledging the separation of the individual economies of household members, there is often a re-assertion of the kinship sentiment that the parties hold for each other: '... oh, but we always help each other. If I finish first, I help my mother, or she helps me. Sometimes my father does not want to accept his fee for selling the cakes. He says he has enough for his cigarettes.'

Another example will show the tensions and ambivalence experienced in a family involved in more complex business arrangements. The case was described by a daughter:

When my father was trying to start his business, some of his siblings gave him some money, maybe $50 each, to help him. He did not ask, they just wanted to help. Now my mother says they were trying to build up obligation, and she feels they are 'bringing up old matters behind our backs' (*bangkit*), to show that they have done us a favour. Really they are envious of my father's success and we are not on good terms with them now. My father wants to tender for a coffee shop, and rent out some of the stalls to his married children. It is taken for granted that each will operate separately, to avoid problems over the sharing of the proceeds. Otherwise some would say their food was better and more popular and they could make more money on their own. There will be problems anyway, because those who see the other stalls doing better than theirs are bound to feel bad about it. My father lent my brother one of his stalls to help him pay off some debts from a previous joint venture which failed. My brother does give him some money to cover the expense of utilities, but the amount he should give has not been specified, nor whether he should compensate my father for his loss of business. My brother has put all my father's stock aside and sells his own goods, so that he can give the stall back as it was with no *bangkit*. But because of this loan, my brother's wife feels obliged to give all of us bigger gifts of money at Hari Raya [major Muslim festival celebrating the end of the fasting month]. She feels the situation is awkward.

This is an example of the sort of entanglements and complications that Malays find uncomfortable and troublesome between close kin. It is clear to everyone that calculations of self-interest are being made, and yet it is not possible, among kin, to clarify these calculations and say clearly 'business is business'.

A third example of a type of economic venture that closely involves family members is a wedding. A wedding has other purposes than simply making money, but its economic aspects are discussed here because they illustrate the co-operation of kin in a short-term project, and the basis of the division of the proceeds. One parent, usually the mother, becomes the head of the event. She supplies most of the capital from her own savings. Her husband may supply part of the capital, either on the basis of a partnership, or on the less explicit basis of joint effort or 'helping out'. Because this is an occasion where a large amount of money will be required, the mother is able to mobilize her relationships with her husband and other children by asking them to 'help' her with cash. They usually do this on the basis of a final gift, not expecting any financial return for their money. If relations are a little strained, the children may still make this contribution but comment out of earshot that their mother is clever at making money from other people's capital. Other kin, and all invited guests, give money as they attend the event. This is seen as part of a cycle of balanced reciprocity where people return the cash that had been given on the occasion of other weddings the hosts had attended, or will give at weddings to come. Kin give their contributions to either husband or wife, and other male guests give to the male host, and females to the female host. This is done discreetly in the handshake as guests leave the wedding. Since husband and wife often attend weddings separately, each going to those of their own friends and acquaintances, the amount of money each receives when they themselves host a wedding depends partly on how good and widespread are their relations with others, and how diligent they have been in attending weddings in the past.

After the event, there is often a surplus over the capital laid out, which is one incentive to hold very large weddings and expand the number of kin and friends in the circle of reciprocity. It is always a gamble whether the thousand or two thousand guests for whom food has been prepared will actually attend, and for this reason as well as the inappropriateness of making business out of a wedding, there has been much criticism of these practices by Muslim and Malay organizations and by some members of the Malay public. The hosts nearly always deny that a profit was made, or say they made just a little to cover their expenditure of effort (*duit penat*). If husband and wife regarded the capital laid out as a partnership, they each take back their share, either by keeping whatever money the guests gave to them or by pooling the money and redividing it according to the original capital outlay. If the husband regarded his contribution as a gift to the wife, or 'helping out', he may be prepared to let her keep all the proceeds.

It has now become common for a reception to be held separately from the main wedding feast. This is a smaller affair of 100–200 guests, and it is the occasion to which the couple invite their own colleagues and other prestigious acquaintances, both Malay and non-Malay. It is more lucrative than the main feast, as the younger guests have more money at their disposal so the cash contributions and gifts are usually more substantial. The gifts received are often a return for gifts the couple themselves had made when they attended the weddings of their peers, and they form a separate circle of reciprocity from that in which their parents are involved. If the couple finance the feast themselves, they receive all of the proceeds. If a parent provides the capital, there are three possible arrangements: all the proceeds are given to the financier, as the profit on his capital; the capital is repaid and the balance, mostly gifts in kind, is kept by the couple; or the parent makes a gift of the proceeds to the child. As one father stated, 'I paid for the food, but her friends gave the money to her hand, not to mine; so I lost out, she gained.' Though he claimed to have made a gift, the father phrased the financing of the wedding in terms of a business deal in which he was the loser. The statement was made to the researcher, and he would probably not have put the economic facts so clearly in front of his daughter, unless he softened their harshness by a teasing or joking manner. As he valued his relationship to his daughter, he would not want to make his economic calculations explicit.

This discussion of the financial arrangements for weddings shows how a wedding becomes the private enterprise of one family member. Others may contribute help in cash or kind, but the entrepreneurial aspect of the wedding is organized as a private, individual, rather than a joint family, venture. At the end of an exhausting day's work, children could be heard grumbling that they had been in favour of a small celebration, but the mother had 'wanted to do business' so, to avoid public embarrassment, the children had felt obliged to help.

Part II
Structuring Practices:
Cultural and Economic Differentiation
in Singapore 1959-1984

Introduction to Part II

PART I analysed the inner workings of the Malay household. It described the cultural terms in which labour services and cash are brought together to maintain the current household, and to sustain the young and old across the various stages of the life cycle. In order to understand how contemporary Malay householding relations came to be organized in their current form, reference was made to the Malay cultural heritage and deeply embedded cultural ideas about the individual, and the way relations between individuals can and should be constructed and maintained. The current material conditions in Singapore, such as the commodification of goods, services, and time, the availability of work by age and sex, provision for pensions, and housing, were examined in order to understand the framework of practical opportunities and constraints within which household life is pursued. Attention was focused on the minutiae of daily transactions in the household, to present an account of the way in which cultural knowledge is applied to changing economic conditions, making the present Malay household form a unique, historical creation and not the predetermined outcome of abstract cultural ideas or material conditions.

Analysis of the contemporary Malay household found that the way householding relationships are currently organized tends to restrict the motivation and ability of Malays to accumulate material wealth through entrepreneurship. The incomes of household members are not pooled, there is little incentive to pass on material wealth to descendants after death, and there is no expectation of achieving social mobility for the whole family, parents and siblings, through the educational, entrepreneurial, or other economic achievements of one family member.

Examination of the interaction of economic and cultural factors continues in Part II, though the emphasis shifts from intensive cultural analysis to consideration of the structuring effect of social practices, such as those within the household and the community, on three important aspects of Singapore's national, social, and economic framework: the emergence of socio-economic classes, the growth and enhancement of ethnic group boundaries, and the growing economic gap between the Malay and Chinese communities.

Although a conventional ethnography would perhaps focus on just one community, here it is argued that consideration of some cultural and economic aspects of the Chinese household and community is critical to the understanding of the position of Malays in Singapore. Malays are a minority, comprising 15 per cent of the population while the Chinese are 77 per cent (1980 census). The national economic context in which Singapore Malays seek opportunities to gain a livelihood is structured by Chinese cultural practices (their household form, their penchant for business enterprise) and also by the sheer numerical presence of the Chinese, raising issues of interethnic relations and discrimination. Many practices of the Chinese have an impact on Singapore Malay culture since there is a tendency for the ethnic boundary between the two groups to focus attention on existing cultural differences or to engender new ones which increase the distinctiveness of 'us' from 'them' (a phenomenon described as 'cultural involution' by Benjamin (1976: 122)).

Chapter 6 provides a brief account of the history of the Malay community in Singapore, describing who the Malays are, when and why they came to be in Singapore, and the economic and social aspects of their migration and settlement. It describes the early emergence of ethnic divisions and social classes, and the reasons for the association of Malays and Chinese with different spheres of economic activity. Chapter 7 uses quantitative data, mainly from national censuses and surveys, together with discussion of economic and educational policies, to identify trends towards class differentiation, and the diverging practices and economic and educational performance of ethnic groups. Quantitative data help to uncover some of the unplanned but real, cumulative consequences of practices that take place at the level of day-to-day life in households and communities. Subsequent chapters focus on the social practices within the Malay community that contribute to class formation, and on the differences between the ways that Malays and Chinese have applied their labour to the economy, which have contributed to ethnic differentiation and to the emergence of a significant income gap between the two communities.

6 The Formation of the Singapore Malay Community

The Pattern of Migration and Settlement

IN the 1980 Singapore census, 351,508 people were enumerated as Malays. They formed 14.6 per cent of the population, while Chinese formed 76.9 per cent and Indians 6.4 per cent (Census 1980: II.11).[1] The migration history of the Singapore Malay population is not accurately known, because movements of indigenes between the present-day political entities, Malaysia, Indonesia, and Singapore, were largely uncontrolled and unrecorded before 1965 (Saw, 1970: 21, 55; Census 1970: 47). People indigenous to the South-East Asian area tended to adopt the 'Malay' label for census purposes after migration to Singapore, and this makes the census an inaccurate means of tracing their specific origins. Here, the term 'Malay' is used to refer to the broad census, category, and the following discussion gives an account of the origins of the main subgroups that make up the contemporary Singapore Malay population.

One component of the Singapore Malay population can be identified by the term 'indigenous Malay'. This group comprises the sea-nomads, royalty, and court personnel who numbered a few hundred at the time of Raffles' arrival in Singapore in 1819 (Turnbull, 1977: 5).[2] It includes subsequent migrants from Malaysia and the Riau islands who claim indigenous Malay origins in contrast to the other major subgroups, which are the Javanese and Baweanese. Hundreds of indigenous Malays came to Singapore from Malacca within the first few months of its founding by Raffles, in response to a British appeal for settlers (Siebel, 1961: 27). In subsequent years, economic opportunities opening up in Singapore attracted more indigenous Malay migrants from Riau, Penang, Malacca, and Johore (Roff, 1967: 33; Census 1931: 72). The great majority of the indigenous Malay population settled in urban areas, and pursued urban occupations (Siebel, 1961: 35). In the 1931 census, 11,290 indigenous Malay males were recorded as gainfully occupied, and of these, 2,075 or 18 per cent were fishermen, and 1,401 or 12 per cent were farmers (Census 1931: 262). In total, only 30 per cent of the indigenous Malay working population in 1931 were employed in the rural sector, and it is important to note this since many misapprehensions have arisen from the mistaken view that indigenous Malays or the Malay population as a whole were predominantly involved in rural pursuits during Singapore's pre-1959 history.

Another major component of the indigenous Malay population were young men recruited from Malaysia into the British uniformed services, especially from the 1930s to the 1950s. In 1957 there were over 10,000 Malays employed in these services and most of them were indigenous Malays, since the British preferred not to recruit Malays of Javanese or other Indonesian origins (Betts, 1975: 41; Djamour, 1959: 5). Most of the servicemen tended to return to Malaysia upon retirement or completion of their contract, so they may not form a very large proportion of the present Singapore Malay population. According to the 1970 census, 25,000 Malaysian-born Malays returned to reside in Malaysia during the period 1957-70 (Census 1970: 263) and it is likely that many of these returned migrants were ex-servicemen and their families.

The second major subgroup of the Singapore Malay population is the Javanese. They were found to number 16,063 in the 1931 census (Siebel, 1961: 18, 19). The 1980 census records that only 6 per cent of the Malay population identified themselves as belonging to the Javanese subgroup, but personal data gathered in Singapore indicate that at least 50-60 per cent of the present Singapore Malay population is of Javanese ancestry.

The Javanese came to Singapore in several distinct waves. There were craftsmen and merchants who established a trading centre in the area of the city known as Kampong Java in the mid-nineteenth century. Their crafts included metal- and leather-work, and they traded in cloth, spices, religious texts, and other goods. Pilgrim brokers were among this mercantile group, and they played a key role in promoting Javanese migration. Dutch restrictions on the pilgrimage to Mecca caused between 2,000 and 7,000 Javanese annually to make their pilgrimage via Singapore between the mid-nineteenth century and 1910, when the restrictions were eased (Roff, 1967: 39). Often, the pilgrims worked in Singapore for months or years before or after their pilgrimage to accumulate the necessary funds or to pay off debts to the pilgrim broker. Many stayed permanently in Singapore, forming the nucleus of an urban, literate, Muslim community (Roff, 1967: 43).

Other Javanese migrants came to Singapore with the help of the pilgrim broker. Some were independent migrants, usually bachelors, who paid their passage and lived in the broker's house on arrival until they established themselves in a trade (often peddling cooked food) or obtained gardening or other labouring jobs which provided living quarters. The pilgrim brokers also recruited bonded labourers known as *orang tebusan* who worked for Malay or Javanese employers and helped to clear land for settlement in nearby Johore (Roff, 1967: 37; Tunku Shamsul Bahrin, 1967a: 237). The practice of bonded labour continued at least until the 1920s. It is estimated that between 1886 and 1890, 21,000 Javanese labourers signed contracts with the Singapore Chinese Protectorate and were sent out to areas of labour demand. In the first decade of the twentieth century the demand was for rubber-estate labour.

Many stayed on after the expiry of their contracts and opened up land to settle in Johore (Roff, 1967: 36-8).[3]

The Javanese component of the Singapore population was augmented after the Second World War by two movements, neither of which is well recorded. The first was Javanese conscript labour brought by the Japanese to Singapore, estimated to number 10,000 (Turnbull, 1977: 216). The second was an indirect migration of Javanese to Singapore via Malaysia. The 1970 statistics on the date of first entry to Singapore show that there were 21,324 Malaysian-born Malays who moved to Singapore in the period 1946-55, and a further 29,679 between 1956 and 1970 (Census 1970: 262-3). Interviews conducted in Singapore indicated that a very large proportion of these post-war migrants were young men or whole families of Javanese descent who were leaving the insecurities of Johore smallholdings to find work in the city. Not all the earlier Javanese settlers had been successful in establishing land claims: many were wage labourers and the whole settlement population was dependent on cash crops with unstable rates of return.[4] Most were uneducated and unskilled and they swelled the lower economic levels of Singapore's post-war Malay society.

Like the Javanese, the Baweanese too came to Singapore in search of urban, waged work. Baweanese migrated to Singapore from the late nineteenth century until after the Second World War. Their migration was provoked not by rural poverty, since Bawean Island had abundant cultivable rice land, but by the search for cash incomes. Their intention in coming to Singapore was to secure gold and consumer goods with which to return to Bawean or to establish themselves permanently in Singapore. Their number, too, is underestimated in the 1980 census, and an official of the Bawean society reckoned that about 70,000 people or 20 per cent of the present Singapore Malay population is of Baweanese descent. They migrated independently, usually as young men, and lived initially in communal houses (*pondok*). Many obtained work as syces and later as drivers, an occupational niche they secured as a group through the practice of mutual recommendations (Abdullah bin M. Baginda, 1967; Vrendenbregt, 1964).[5]

A popular view held by Singapore people about the Singapore Malays is that they form a predominantly indigenous, rural, and unchanging population. Contrary to this image, the brief review of their migration history provided here shows that large numbers migrated to Singapore only in the 1945-70 period, and though most of them came originally from rural backgrounds, both earlier and later settlers came to Singapore in order to seek an urban livelihood. The great majority of Malays were never engaged in rural pursuits in Singapore, and their settlement pattern reflects their urban orientation. Many lived in quarters provided by employers as they worked as servicemen, gardeners, drivers, or employees of government boards such as public works and utilities (a niche occupied by unskilled Malay labourers following the war). Massive squatter settlements, such as Geylang Serai to the east of Singapore, were established to accommodate the post-war influx of Malays from

Malaysia. Other longer-established settlements expanded in population during the 1950s and 1960s as rental accommodation was built for the incoming migrants.

Singapore Malay settlements, known as kampongs, were very different from the integrated rural villages of the Malay hinterland which are known by the same term. The big Singapore kampongs were inhabited by people largely unrelated by kinship, many of whom were renters who frequently moved from one settlement to another. There was a mix of Malay subgroups, and kampong dwellers were engaged in a wide variety of urban occupations. This is important to note, because the use of the same term 'kampong' for both rural Malaysian villages and urban squatter settlements has misled many into thinking that the Singapore settlements were rural, stable, and ancient. In fact they were, with a few exceptions, urban, unstable, and mostly built in the 1950s.

Another key feature of Malay migration to Singapore is its individualistic nature, and this is crucial to the question of ethnic and economic structuring. With the exception of the Javanese bonded labourers, few of the migrants to Singapore or Malaysia worked for other Malays. They migrated as individuals, they paid off whatever debts they had incurred for their passage, and they set about finding work for themselves. Few Malays had established businesses in which they could employ their migrant kin or village mates, and the help they were able to offer new migrants was minimal. In part, this independence was chosen by the migrants themselves, many of whom were young men seeking adventure and escape from the constraints of parental authority. They did not seek out kin or any other possible patrons, since they preferred the egalitarian company of peers, renting a room together and sharing the cost of food or eating outside the home. Single women migrants were often divorcees seeking to escape village and family gossip, and they found work for themselves as domestic servants where, again, accommodation was often provided. It was the relative freedom and anonymity of the city, and the possibility of supporting themselves as independent wage-earners, that attracted these individuals to Singapore.

Though Malays did choose to group together in kampongs for the benefits of religious and community life among their own kind, they did not organize economic ventures among themselves, nor combine their resources in order to penetrate and dominate sectors of the wider Singapore economy. In terms of their occupational pattern, Malays were from the earliest days thoroughly integrated into multi-ethnic Singapore, working for non-Malays.

The migration and occupational pattern of the Chinese was very different. A proportion of the Chinese migrants came to Singapore under large-scale indenture movements, especially in the late nineteenth century. Others were recruited 'voluntarily' but became bound by debts to a labour recruiter, ship captain, or lodging-house keeper in Singapore. The migrant then became a member of a *kongsi* or group of workers under a contractor. The contractor acted as an intermediary between the Chinese workers and the European employers, and was able to retain his control in part because of the constraints of language which prevented direct employment (State of Singapore, 1960: 4). Most British colonials could speak some Malay, but few spoke the varied Chinese dialects of these migrants.

Other Chinese migrants joined kin or quasi-kin, co-villagers, or co-dialect speakers and worked in their businesses under their paternalistic authority and not always in favourable conditions. To better himself, the migrant had to start a business of his own, but particular branches of business and areas of the city were under the protection of Chinese secret societies and subgroup monopolies. Even as an entrepreneur, the Chinese migrant was necessarily integrated into an entirely Chinese world, which both provided opportunities and benefits and also engendered abuses. The pattern of Chinese migration supported a structure of power and authority in the Chinese community, and gave some Chinese direct or indirect control over the labour power of others (Freedman, 1979: 65, 73). It kept whole sectors of the economy under the control of particular groups of Chinese, and totally excluded non-Chinese.

The contrast in the pattern of Malay and Chinese migration described here was the source of early ethnic and class division in the Singapore economy. Economic differentiation, ethnic differentiation, and the important question of Malay and Chinese participation in entrepreneurship are subjects that will be discussed in more detail in the following chapters.

Ethnic Sentiment and the Development of New Cultural Forms

As the various groups of migrants from the Malayo-Indonesian Archipelago came to Singapore, they did not adopt any clearly defined or pre-existing Singapore Malay culture. Singapore Malay culture has developed out of the varied traditions of the migrants, and out of the conditions of life in Singapore. In the course of migration to Singapore, the rural-based 'little traditions' that distinguished the Javanese, Baweanese, and other South-East Asians from each other have gradually eroded, and the great court traditions of Java and Malaysia have lost their relevance. The features that positively characterize Singapore Malayness, the new identity developed *in situ*, are the Malay language and Islam. These are the 'great traditions', literate and urban, which developed to new heights in Singapore and to which all incoming migrants were exposed.

Apart from the rather minimal cultural criteria of Islam, the Malay language, and a willingness to participate in social, religious, and neighbourhood activities with other Malays, the critical factor defining Malay identity and Malay community boundaries has been the development of ethnic sentiment distinguishing Malays from Chinese, Indians, and other non-South-East Asians. While cultural differences existed between Malays and Chinese throughout the colonial era, the various population groups tended to remain separate in their own schools, areas of residence, and occupations. The political and economic circumstances engendering direct competition between Malays and Chinese did not arise until the imminent withdrawal of the British in the 1940s and 1950s, when it became necessary to confront the question of the division of power in the post-colonial era. Ethnic sentiment emerging in this period came to be known as 'Malay nationalism' and Roff has described its development, particularly during the 1940s (Roff, 1967). Other writers have described the events surrounding the merger of Singapore and Malaysia in 1963, and the subsequent separation of Singapore as an independent nation-state in 1965 (Betts, 1975; Bedlington, 1974; Turnbull, 1977). These events greatly heightened ethnic sentiment among Singapore Malays who were made conscious of their isolation as a Malay minority in a state that would be dominated by Chinese ethnic groups and by Chinese political and economic power.

Migration patterns in the colonial era, economic and cultural trends, and the development of ethnic sentiment in the Malay and other communities combined to produce in Singapore the predominant, even unquestioned, popular and official view that the primary divisions in Singapore society have always been ethnic, rather than socio-economic, in nature. The society is viewed as comprising three distinct ethnic blocks, Chinese, Malay, and Indian, which are internally homogeneous (or nearly so), and each of which is associated with particular cultural practices and forms of economic activity. It is against the background of Singapore's general predilection for ethnic thinking (described in more detail by Benjamin, 1976) that the following discussion of the emergence of specific aspects of ethnic and economic differentiation in Singapore since 1959 should be understood.

1. The Indian community has been excluded from detailed discussion here because it is a smaller group numerically and because its inclusion would over-extend and complicate the comparisons to be drawn between the Chinese and the Malay communities. The census uses a fourth population category termed 'other' which comprises people of Arab, European, and other diverse origins.
2. See Pang Keng Fong (1983) and Turnbull (1977) for accounts of the history and intrigues of the Malay Royalty in Singapore. See Normala Manap (1983), Vivienne Wee (1984) and Vivienne Wee (n.d.) for accounts of the indigenous population and their relationship to Malayness.
3. See Ramsay (1956) and Tunku Shamsul Bahrin (1967a, 1967b, 1970) for accounts of Javanese migration and settlement in Johore.
4. See S. Husin Ali (1964, especially pp. 20, 28, 36, 139) for an account of a Javanese village in Johore, its ancestry, land and wealth distribution, and the migration of its youth out of the village. This is the type of village from which the post-war Malay migrants to Singapore came, according to the life histories and family background data collected.
5. Other smaller groups that came to Singapore were the Bugis, the Banjar, and various groups from Sumatra. Shortage of space prevents a discussion of the history and migration of these groups here. They are all included within the census category 'Malay'.

7 Malays in the National Economic and Education System

THIS chapter uses quantitative data, and analysis of government policy, to investigate when and how Malays came to be over-represented at the lower levels of the Singapore economy. The data show that there are differences in the way Malays and Chinese apply their labour to the national economy which have tended to disadvantage Malays when their performance as a group is compared to that of the Chinese. These differences in practices between the two groups are related both to their respective cultural traditions, and to the material opportunities that have been accessible to Malays and Chinese as ethnic groups. At the same time, however, the data on Singapore clearly show that *both* Malays and Chinese are participants in an education and economic system in which advantages are unevenly distributed, and the majority of both groups encounter limited possibilities for socio-economic mobility.

Analysis of statistical data helps to identify some of the direct or indirect ways in which the practices of everyday life in the Malay household and community have affected their participation in the national economy over a period of time, both in relation to the Chinese and in relation to the overall process of the development of socio-economic classes. Later chapters will further investigate the cultural patterns in the Malay, and Chinese, household and community which underlie these statistical trends.

Comparison of the trends in Malay and Chinese participation in the national economy is made possible by the tendency in Singapore to classify social and economic data by ethnic group. This form of classification also poses a danger, in that differences in Malay and Chinese performance in the national economy which reflect their over- or under-representation in certain economic strata, may appear to be *caused* by ethnic or cultural factors simply because the data is presented by ethnic group. This point will be important to the discussion of ideology in Singapore in Chapter 11. The present chapter attempts to separate out the effects of cultural traditions related to the Malay and Chinese heritage from questions of the distribution of economic opportunities and rewards.

The Economic Position of Malays and Chinese Prior to 1959

The preceding chapter showed that Malays were full participants in the urban economy of Singapore from the moment of their arrival on the island. They were excluded from large sections of the economy dominated by the Chinese, but they had an established niche as the employees of Europeans. Prior to 1959, the majority of Malays were not generally worse off economically than the majority of the Chinese. This was shown by Goh Keng Swee, who conducted a social survey of Singapore in 1953-4. He noted (1958: 100) the discrepancy between favourable Malay economic performance and the popular image of Malays as economically backward: 'There is a widespread impression that this group, which consists mostly of Malays born in Singapore or the Federation, forms the economically depressed section of the Singapore community....' He found instead (1958: 19) that 'the local born Malay is shown to have held his own against the other communities. It is only in the sense that comparatively few of them succeed to reach the best paid occupations that one can say that the Malay community has been economically unsuccessful.' His findings (1958: 19) revealed that only 5 per cent of Malays, compared to 16 per cent of Chinese, had household incomes of over $400 per month. For the group with which his survey was concerned, that is the remaining 84-95 per cent of the population, the Malay 'average household income is, in fact, larger than that of the immigrant Chinese, who are supposed to be the most successful and enterprising section of Malaya's population' (Goh, 1958: 100).

Goh recorded household incomes, and also devised a measure of poverty which detected households with an income lower than that needed for subsistence at defined levels. His findings, summarized in Table 7.1, show the favourable economic position of the local-born Malay population in the 1950s. Goh uses the term 'local born' to refer to Malays born in Singapore or Malaysia. 'Immigrant' Malay refers to those born in Indonesia, who made up 25 per cent of the total Malay

TABLE 7.1

Malay and Chinese Monthly Household Incomes and Percentage of Households in Poverty, 1953

	Household Income ($ per month)			*Percentage of Households in Each Commumity in Poverty*
	Lower Quartile	*Median*	*Upper Quartile*	
Local-born Chinese	105	161	240	20
Immigrant Chinese	115	148	216	25
Local-born Malays	118	160	217	20
Immigrant Malays	94	119	166	34

Source: Adapted from Goh, 1958: 99, 135.

category in his study (Goh, 1958: 22). This group was the weakest economically, and life-history data on Singapore Malay origins indicate that a number of them were ex-Japanese conscripts from Java who had been left totally destitute in 1945. Their presence lowered the overall average income of the Malay category in the 1950s.

The educational pattern in 1953-4 is also of interest. Table 7.2 shows that for the majority of the population (excluding the élite 5 per cent of Malays and 16 per cent of Chinese), local-born Malays were more likely to avail themselves of the limited vernacular education facilities available, than were either local- or foreign-born Chinese. A similar proportion of local-born Malays and Chinese obtained at least some English education.

The pre-1959 élite in Singapore, with incomes of more than $400, was made up of entrepreneurs and professionals. The business élite was weakly developed among Malays, for reasons to be discussed in detail in Chapter 9. The professional élite was largely restricted to people who had received their education in English. The Chinese professionals were mainly found among the local-born *Peranakan* who were early settlers in Malaysia and Singapore and who had sent their children to English schools since the late nineteenth century (Roff, 1967: 110; Nagata, 1979: 28). Opportunities for Malays to obtain an English education prior to 1959 were more limited. Colonial policy during the nineteenth and early twentieth centuries reserved English education and government positions for Malay royalty and restricted the access of the Malay masses, since the British feared an over-educated indigenous population. Colonial policy was intended 'to educate the rural [Malay] population in a suitable rural manner to equip them to continue to live a useful, happy rural life' (Roff, 1967: 28). The policy extended to Singapore, despite the fact that Malays there were generally not rural. When opportunities

TABLE 7.2
Educational Level of Malay and Chinese Male Heads of Household, 1953 (per cent)

	Local-born Chinese	Immigrant Chinese	Local-born Malays	Immigrant Malays
No education or less than 2 years	30	42	15	69
Elementary vernacular	36	40	56	19
Secondary vernacular	5	10	3	3
Elementary English	7	–	3	1
Secondary English	14	–	9	1
English and vernacular	5	1	12	4
Unknown	3	6	1	3
	100	100	100	100

Source: Adapted from Goh, 1958: 40.

for English education did become available to some Singapore Malays able and willing to finance their children in Christian Mission schools, fear of conversion was a further impediment. These factors resulted in the Malays being left at least a generation behind the Chinese in establishing their English-educated clerical and professional classes. These two factors, lack of participation in business and lack of English education, account for the presence of only 5 per cent of Malays compared with 16 per cent of Chinese in the high-income group in 1953.

It is the under-representation of Malays at the élite level which has generally been the focus of popular, media, government, and academic attention, and this has tended to obscure the relatively equal position of the great majority of the Malay and Chinese population prior to 1959. In both ethnic communities, the gulf between the vernacular- and English-educated, was extreme. For the Malays, the 'two diverging (Malay) vernacular and English educational systems produced two distinct classes, each culturally, intellectually and economically divorced from the other', while for the Chinese, 'the growth during these four decades (1900-41) of both Chinese and English education created and consolidated two solid and distinct classes' (Chai Hon Chan and Yong Ching Fatt cited in Gopinathan, 1974: 3). The standards in the Chinese and Malay vernacular schools were equally low, and the vast majority of Malay and Chinese children before the Second World War received at best primary education, and went into manual jobs (Turnbull, 1977: 146).

Chinese and Malay Economic Positions 1959-1980

The available census data show that the position of the Malays as a group relative to the Chinese has deteriorated since 1959. This is shown by the occupational distribution and by income figures in Tables 7.3 and 7.4.

While the gap between the percentage of Malays and Chinese in the

TABLE 7.3
Malay and Chinese Working Males Aged 10 Years and Over by Occupation, 1957, 1970, and 1980 (per cent)

	1957		1970		1980	
	Malays	Chinese	Malays	Chinese	Malays	Chinese
Professional and technical	2.8	3.6	4.4	6.7	4.6	8.3
Administrative and managerial	0.3	1.8	0.3	2.1	0.8	6.7
Clerical	15.0	11.5	13.9	11.2	13.0	8.5
Sales	3.0	23.8	3.7	20.0	3.1	15.4
Services	13.7	9.0	21.6	7.6	18.0	6.2
Agriculture and fishing	10.0	7.6	5.9	4.6	3.1	2.5
Production and transport	42.1	42.3	45.4	43.2	53.5	43.2
Not classifiable	13.1	0.4	4.8	4.6	3.7	9.1
	100.0	100.0	100.0	100.0	100.0	100.0

Sources: Derived from Census 1970: 277, 285; Census 1980: IV.66.

two higher occupational categories in 1957 was 2.3 per cent, by 1970 it was 4.1 per cent and by 1980, 9.6 per cent. The 1980 census shows that within the professional category, Malays were concentrated at the lower levels as teachers and technicians (Census 1980: IV.142, 144). In 1957, Malays and Chinese had the same proportion of their population in the lower manual category, but by 1980 there were 10 per cent more Malays in that category.

Income figures give a clear indication of the trend towards a relative decline in the Malay position, and suggest that the decline was most marked in the late 1970s. In 1966, 47.7 per cent of Malays earned less than $150 per month compared with 43.5 per cent of all races, which suggests that Malays were holding their own in the 1960s, at least at the lower end of the income scale (Aljunied, 1980: 82). Figures for income by ethnic group were made available for the six years 1975-80, and they show a marked decline in the Malay position during the period, affecting both the majority at lower-income levels and the élite. The discussion and tables below are restricted to male workers because the entry of large numbers of Malay females into low-paying manufacturing jobs in the late 1970s would otherwise distort the comparison over the years.

TABLE 7.4
Employed Male Malays and Chinese by Monthly Income, 1975, 1978, and 1980 (per cent)

	Earning Less than $400 per Month		Earning More than $1,000 per Month	
	Malay	Chinese	Malay	Chinese
1975	62.6	67.1	0.8	7.0
1978	75.8	53.3	1.8	8.9
1980	64.1	41.8	2.7	12.9

Sources: *Report on the Labour Force Survey*, 1975: 95; 1978: 98; 1980: 69.

Table 7.4 shows that while the percentage of Malays in the lower-income group increased, 25.3 per cent of Chinese male workers climbed out of that lower-income group between 1975 and 1980. The Malay position declined relative to the Chinese and also absolutely, as the rise in the consumer price index during the period meant that the $400 earned by 64 per cent of the Malay population in 1980 was worth approximately 17 per cent less in 1980 than it had been worth in 1975 (*Economic and Social Statistics*, 1960-82: 213).

Budgets obtained from lower-income households in 1982 indicated that $400 could not sustain a family of four for a month, so all those Malay and Chinese households relying on a single male income-earner were in poverty. From this income of $400, approximately $100 would be deducted for the compulsory savings scheme; $70 was needed for the rent and utilities of the smallest one-room flat; $15 paid the school fees of two children, assuming they walked to school; and at least $40 was needed for the wage-earner's transport. This left $175, or less than $6 per day, for all the food consumed by the family, including the wage-earner's consumption at work and the children's consumption at school, leaving no margin for emergencies, medical expenses, clothing, furnishing, or school books.[1]

The following sections examine the reasons for the decline in the Malay economic position relative to the Chinese, and their absolute economic decline during the 1959-80 period. The issues of labour-force participation, entrepreneurship, discrimination, national economic restructuring, and education are discussed.

Labour-force Participation and Household Size

If in 1980 the individual wage earned by over 60 per cent of Malay men and 40 per cent of Chinese men was insufficient to support a family, then the number of income-earners per household and household size are clearly crucial factors in the survival, welfare, and potential for economic advancement of Singapore households. Malay households were worse off than Chinese households throughout the 1960s and 1970s because of a lower female labour-force participation rate, and a larger household size with a higher rate of dependency.

Table 7.5 shows that Malay women were slower than Chinese women to join the formal labour force. This meant that fewer Malay households had the benefit of the extra wages earned by a wife or daughter during the 1960s and 1970s, although the labour-force participation rate of Malay and Chinese women had equalized by 1980. Fifty-four per cent of Singapore households had more than one working person in 1980, though these figures are not given separately by ethnic group (Census 1980: VII.8).

The most prominent reason for the relative lack of Malay female labour-force participation during the 1960s and early 1970s was the absence of work opportunities. While Chinese females could be employed in various capacities in the Chinese-speaking trade and manufacture sector, it was the growth in manufacturing in the multinational sector in the late 1970s that first provided mass employment opportunities for

TABLE 7.5
Malay and Chinese Female Economic Activity Rate by Age, 1957, 1970, and 1980 (per cent)

Age	1957		1970		1980	
	Malay	Chinese	Malay	Chinese	Malay	Chinese
15-19	5.4	33.8	27.7	45.9	55.5	50.3
20-24	5.7	30.3	27.6	59.3	73.1	80.2
25-29	6.6	13.7	18.5	34.7	54.8	60.0
30-34	8.3	9.7	19.0	25.9	43.7	42.0
35-39	10.3	8.5	22.7	21.8	37.5	
40-44	10.8	9.0	28.7	19.7	25.6	31.7
45-49	9.5	8.8	33.1	19.0	18.7	

Sources: Chang et al., 1980: 52; and derived from Census 1980: IV.41.

Malay women (Cheng, 1980: 31). This sector of the economy is scheduled to be displaced by high-technology, skill-intensive industries under the government's economic plan for the 1980s and beyond, and it is likely that there will be many redundancies amongst female workers. Since Malay women are concentrated in the multinational manufacturing sector, they are more vulnerable to the decline of this particular sector than Chinese women, who are involved in a wider range of occupations.

TABLE 7.6
Malay and Chinese Working Females Aged 10 Years and Over by Occupation, 1980 (per cent)

	Malays	Chinese
Professional and technical	5.3	10.5
Administrative and managerial	0.3	2.5
Clerical	17.5	29.6
Sales	4.5	11.7
Services	15.9	12.9
Agriculture and fishing	0.8	1.0
Production and transport	55.6	30.7
Not classifiable	0.2	1.1
	100.0	100.0

Source: Census 1980: IV.66.

A later chapter will discuss the Chinese household arrangements that encourage female labour-force participation, especially among unmarried girls. It will also suggest that Chinese women workers are likely to put a higher proportion of their wage at the disposal of their families than Malay women do. These are differences stemming from within the household that have had an effect on the relative economic standing of Malay and Chinese households.

Household size was another important factor that affected the relative position of Malay and Chinese households during the 1960s and 1970s. Whereas the fertility rate of Malays and Chinese was similar in 1957, the Chinese rate dropped very rapidly during the early 1960s while the Malay rate did not decline until the late 1960s, reaching a par with the Chinese

TABLE 7.7
Malay and Chinese Total Fertility Rate, 1957-1976

	Malays	Chinese
1957	6.3	6.5
1960	6.6	5.7
1965	6.3	4.3
1970	3.5	3.0
1973	2.8	2.8
1976	1.9	2.2

Source: Adapted from Chang et al., 1980: 59.

rate by 1973. As a result of the higher fertility rate, Malays had a higher dependency ratio than Chinese during the 1960s and 1970s. The average size of households comprising one nuclear family was 4.9 for Malays and 5.4 for Chinese in 1957, and 6.1 for Malays and 5.6 for Chinese in 1970 (Chang et al., 1980: 68). The reason for this difference in fertility rate is probably related to the later age at marriage and higher labour-force participation rate before and after marriage of Chinese women compared to Malays. Age at marriage and labour-force participation are in turn related to economic opportunities and to differences between Malay and Chinese household forms, which are explored in detail later in this chapter and in Chapter 8.

The correlation between socio-economic factors and fertility should also be noted, since education and income affect the fertility rate of all ethnic groups. Age is a related factor, as it is the younger Malay and Chinese women, who are better educated and have higher household incomes than their parents, who tend to have fewer children (Chang et al., 1980: 80). The 1980 census does not provide a correlation of household income with fertility, but the correlation of fertility with female education gives an indication of the effects of socio-economic status:

TABLE 7.8

Malay and Chinese Females who have ever been Married, by Highest Qualification and Mean Number of Children Born Alive

	Malays	Chinese
No qualifications	4.91	4.27
Primary	2.43	2.28
Secondary	1.42	1.56
Tertiary	0.70	1.52

Source: Census 1980: IX.70.

Entrepreneurship

National statistics show clearly that while few Malays are involved in entrepreneurship, this is an important form of economic activity among the Singapore Chinese:

TABLE 7.9

Malay and Chinese Working Males Aged 10 Years and Over by Employment Status, 1957, 1970, and 1980 (per cent)

	1957		1970		1980	
	Malays	Chinese	Malays	Chinese	Malays	Chinese
Employee	93.1	67.4	93.5	70.2	95.3	75.2
Employer	0.2	5.1	0.7	3.9	0.5	6.3
Own account worker	6.2	22.7	5.9	22.3	3.9	16.2
Unpaid family worker	0.5	4.8	0.3	3.6	0.4	2.3
	100.0	100.0	100.0	100.0	100.0	100.0

Sources: Census 1970: 284-5; Census 1980: IV.34.

Only 4-7 per cent of Malay males in the work-force between 1957 and 1980 have been either employers or own account workers, while the Chinese participation rate in these two categories has been in the range of 22-28 per cent. There was a decline in the percentage of Chinese in the own account worker category from almost 23 per cent in 1957 to 16 per cent in 1980, but data concerning the returns on entrepreneurship indicate that small-business activity has become much more lucrative over the years. In Goh's survey of 1953, 'own account workers' earned only a few dollars more than employees, so for the mass of the population (excluding the top income-earners), small-scale entrepreneurship did not generally provide an improved standard of living (Goh, 1958: 100). By 1980, when incomes are related to educational level, which is the key factor in determining the income of employees, it is evident that entrepreneurship (designated in this part of the census by the term 'self employment') had become a very significant vehicle for the attainment of higher incomes and upward mobility:

TABLE 7.10
Income of Males Aged 10 Years and Over by Education and Employment Status, 1980

	Percentage Earning Less than $500 per Month		Mean Monthly Income, in $	
	Employees	Self-employed	Employees	Self-employed
No qualifications	74.8	53.3	410	558
Primary	66.6	40.0	476	704
Secondary	36.7	20.2	942	1,426

Source: Census 1980: VII.46-7.

The participation of a higher percentage of Chinese in entrepreneurship and the favourable returns on business activity, even for the uneducated, especially during the 1970s, was a factor which enabled some Chinese to achieve upward mobility and higher incomes. Differential participation in entrepreneurship partly accounts for the decline of the Malay economic position relative to the Chinese position during the period. Although no figures are available, it is very likely that Chinese greatly outnumber Malays in the informal, undeclared private enterprise sector of the Singapore economy, just as they do in formally declared entrepreneurship. This would mean that a number of Chinese households actually have higher incomes than are shown in census data, further increasing the income gap between Malays and Chinese.

Apart from the question of individual incomes derived from entrepreneurship, the involvement of Chinese in small business affects the overall shape of the Singapore economy and the distribution of opportunities for employment. According to a 1977 survey, the commerce sector is the largest sector in the economy. It engaged nearly a quarter of the work-force and contributed 27 per cent of GDP in 1977 (*Report on the Survey of Wholesale and Retail Trades, Restaurants and Hotels, 1977*).

Another source indicates that there was over 100 per cent growth in the small-business sector of the Singapore economy during the 1970s:

TABLE 7.11

Number of Establishments (Commerce and Manufacture) by Size (Number of Workers), 1970 and 1980

Size (No. of Workers)	1970 No.	1970 Per Cent	1980 No.	1980 Per Cent
1-4	20,522	63.1	47,955	69.1
5-19	9,382	28.8	16,424	23.7
20-49	1,561	4.8	3,035	4.4
50 and over	1,064	3.3	1,960	2.8
	32,529	100.0	69,374	100.0

Sources: Yearbook of Labour Statistics, 1970, 1980.

The small-business sector is largely composed of family firms. In 1980, 52 per cent of commercial and manufacturing establishments were sole proprietorships (*Yearbook of Labour Statistics, 1980*). In the commerce sector, 23 per cent of all those employed in the sector in 1981 were working proprietors, partners, or unpaid family workers (*Report on the Survey of Wholesale and Retail Trades, Restaurants and Hotels*, 1981). Workers tend to be recruited on the basis of family ties or on the basis of language. Since most business in this sector is conducted in Chinese dialects, there are virtually no opportunities for Malay employees in this large sector of the economy. The activities of Chinese entrepreneurs serve to enhance not only their own individual earning capacities, but the employment opportunities of the whole Chinese group relative to the Malays.

Discrimination

There has been discrimination both for and against Malays during Singapore's history, and this has been a factor affecting their position as a group relative to the Chinese. The Malays were positively favoured as employees by the British, particularly in the uniformed services (army, police, fire) and in some related clerical, transport, and personal services. In 1957, almost 20 per cent of Malay working men were employed in the uniformed services (Betts, 1975: 41 and Census 1970: 198). The favoured position enjoyed by Malays was lost with the withdrawal of the British defence establishment between 1970 and 1975. The Singapore government did not take over, or renew, the contracts of Malay uniformed personnel, and this is probably a major factor in the decline in Malay incomes during the 1970s. Large numbers of Malays who had built up their pay through years of seniority in the uniformed services were thrown on to the job market with little formal education and irrelevant skills in 1975, just as Singapore was experiencing a recession.

From 1965 onwards, the government of independent Singapore had an unspoken, but widely known, policy of excluding Malays from recruitment into the new Singapore armed forces and police. This was discrimination against Malays, denying them employment opportunities in their major field of expertise. The government felt the policy to be justified on the basis of security considerations, since it believed that Singapore, as a majority Chinese city-state, was vulnerable in the middle of Malay-Muslim South-East Asia. It was presumed that a predominantly Malay security force could not be relied upon to protect Singapore's interests in the event of internal racial strife or external aggression. This made the recruitment of Chinese to the armed forces, and the ousting and exclusion of Malays, an urgent matter of national security. Malay youth were not called up for National Service during the 1970s, and some were still not being called up in 1984. Those who were called up claimed to be placed only in menial capacities, and always excluded from the airforce, commando, and tank units which are the key units in Singapore's defence system.

There was an unfortunate side effect to the non-recruitment of Malays into National Service. Employers in Singapore are generally unwilling to recruit or train young male workers who have not completed National Service or obtained exemption papers as these youths can be called up at any time. Since Malays were not officially exempted from National Service, Malay youths were unable to obtain apprenticeships or regular jobs, and many were forced into an extended limbo period of about ten years from ages 14 to 24 when they could obtain only casual irregular work. Malay organizations protested against this policy, as it was felt that the irregular life-style forced upon these youths was in part responsible for the high percentage of Malay youths who became involved in heroin abuse during the late 1970s. These individuals' long-term employment prospects have been badly affected by their lack of training, irregular work experience, and in some cases drug and criminal records (the government, for example, does not employ ex-addicts, *Parliamentary Debates* 20.3.1979: 929, Ong Leong Boon). Whatever its relation to the heroin problem, numerous personal interviews suggested that the exclusion of Malay boys from National Service during the 1970s and 1980s had a deeply alienating effect on a generation of Malay youths and their parents, who became convinced that Malays were not trusted and not really wanted, by the government of independent Singapore.[2]

Malays believe that discrimination against them by Chinese employers is very widespread or even universal. It is not illegal to specify the race of the employee required in a job advertisement, and advertisements such as 'Malay driver' or 'Chinese secretary' can be seen in the press (Aljunied, 1980: 71). Two employment agencies interviewed in the course of this research stated that employers almost always state the race of the employee they wish to hire, either verbally or on the application form. These agents, one Indian and one Malay, found that discrimination did not hinder their own business with the Chinese community, where they won clients on the basis of their efficiency in providing a specified service. They did find that discrimination affected the employment prospects of Malay secretaries or other office staff who were not acceptable to a Chinese employer with whom they would have to work in daily close contact. This was the case even when the business of the office was conducted entirely in English, so that the Malay applicant's inability to speak Chinese was not a relevant factor. In most cases, the employer's prejudice, as expressed to the employment agent, was general and racial, and not related to specific skills.

Nearly all of the working histories collected from Singapore Malays gave detailed examples of practices among Chinese employers which the Malays believed to be discriminatory. This included practices in recruitment, pay and conditions, and opportunities for training and career advancement. A man compared, for example, the number of years it took for Chinese and Malay workers with secondary-level qualifications to obtain promotion to supervisor level in his factory. One girl described how she and a group of friends were told by a Chinese personnel officer that there were no vacancies in a factory, but then applied again directly to the European manager the following day and were given jobs. Those who wished to obtain training in trades, such as vehicle mechanics, which are organized on the basis of small Chinese-owned workshops, described an experience of frustration as they were obliged to remain at the lowest, menial levels of the trade, and were not accepted as apprentices by the Chinese mechanics. For many Malay workers, the decision to continue their employment in one factory or office for most of their working careers, which might appear to the casual observer to be unambitious, is related to their conviction that they will face difficulties finding other jobs, not because of their lack of qualifications but because of racial prejudice.

Discrimination in employment is based partly on Chinese perceptions of Malay ambitions and capacities.[3] In his study of Malay and Chinese workers in large Singapore establishments, Deyo (1983: 223) found that there is no basis for the belief that Chinese workers are more materialistic, hard-working, or ambitious than Malay workers:

> Among white-collar male workers, Chinese appear to be relatively more satisfied with their current positions and less concerned about advancement. And while blue-collar [Chinese] workers in Singapore *rank* material gains higher in absolute terms, they show lower job involvement, a lesser tendency to interact with supervisors on the job, and less concern with most specific job factors, including pay, benefits, raises, and promotions [than Malays].

Such evidence notwithstanding, there is a widespread conviction among Singapore Chinese that Malays are lazy or, put more charitably, that Malays are interested in spiritual, artistic, or social pursuits but relatively uninterested in material gain. This assessment of the Malay character is evidently not confined to the older generation, as Chinese youths in a study by Leong Choon Cheong expressed similar attitudes: 'Malays are always stupider' (Leong Choon Cheong, 1978: 43); 'Malays are lazy and would stay away from work after they had earned enough to last them for a short period of time' (Leong, 1978: 43); Malays are 'unhygienic, uneducated, unintelligent and culturally lazy' (Leong, 1978: 107); they are 'the laziest and most aggressive people in the world' (Leong, 1978: 92); and 'they're so lazy! Today get pay, tomorrow no work. My former boss at the welders told me that Malays are just born like that–as long as they have enough to eat, they don't want to work' (Leong, 1978: 142). The youths who expressed these opinions are among the 50–60 per cent of the young Chinese population with low education and poor future prospects, so their opinion of Malays is not related to their own manifest abilities, diligence, or success. A later chapter will discuss some of the sources of these views about Malays, which stem from historical circumstances, from real differences in economic behaviour between Malays and Chinese at the level of daily life such as involvement in business enterprise, and from processes related to ethnicity and ethnic boundaries.

Although its effects are not quantifiable, discrimination against Malays in the various forms described here has affected their economic standing relative to the Chinese, particularly in the years since 1959 when Chinese replaced the British in all spheres of political and economic power. Despondency among Malays about the possibility of advancement in Singapore has caused some migration to Malaysia, particularly among Malay professionals or would-be entrepreneurs for whom opportunities in Malaysia are better. For ordinary Malay workers, however, the higher pay available in Singapore continues to make Singapore more attractive, and they note that in Malaysia, too, the employer is often Chinese.

The Malaysian government has a programme of positive discrimination in favour of its Malay citizens but few Singapore Malays are attracted to Malaysia by these advantages. They generally feel that open competition is preferable to discrimination, even when the discrimination is in favour of Malays, and they are sometimes disdainful of the corruption and meagre progress they observe among Malays in Malaysia, who do not appear to have made full use of the opportunities provided under government programmes. Due to the circumstances of Singapore's separation from the Malaysian Federation in 1965, a clause was included in the Singapore Constitution which gives Malays a 'special status' as the indigenous people of Singapore. The People's Action Party government, which has been in power in Singapore continuously since 1959, has always been opposed to the principle of positive discrimination on the basis of race, and has made only one concession to Singapore Malays in granting them exemption from paying full school fees. Interview data indicated that Singapore Malays agree with the principle that they should compete on a non-racial basis with all other Singaporeans, although they are convinced that there is actually discrimination against them which gives them a less than equal chance.

The Distribution of Opportunities and Rewards in the National Economy

The factors relating to labour-force participation, household size, participation in entrepreneurship, the British withdrawal, and the minority position of the Malay community in Singapore, discussed above, account for the decline in the economic position of Malays as a group relative to the Chinese during the period 1959-80. Once the Malays became over-represented in the lower-income stratum in Singapore, their position as a group was subject to further decline relative to the Chinese because proportionately more Malays encountered the set of disadvantages affecting all people in Singapore on low incomes. The distribution of opportunities and rewards in the national economy and education system directly affects the material circumstances of all Singapore households, and indirectly affects the relative economic position of ethnic groups in Singapore.

Although Singapore achieved a high rate of economic growth during the period 1959-80, several analysts have noted that the distribution of the rewards of the expanding economy tended to be unequal. A study of poverty in Singapore was carried out in 1973, using the same stringent standards as those established by Goh to measure the extent of poverty in 1953 (Cheah, 1977; Goh, 1958). The study established that the incidence of poverty had increased during the twenty-year period, and 37 per cent of Singapore households in 1973 did not have incomes sufficient for their basic needs:

TABLE 7.12
Estimates of Poverty, 1954 and 1973 (per cent)

	Standard A (excluding education costs)		Standard B (including education costs)	
	Households	Persons	Households	Persons
1954	19	25	21	n.a
1973	33	34	37	39

Source: Cheah, 1977: 41.
n.a.–Not available.

Cheah (1977: 43) describes this finding as a 'counter-trend towards an increase in poverty among certain segments of the population which overall statistics tend to mask'. Cheah found that poverty was not a phenomenon restricted to widows or the elderly, or households handicapped in some way, but the majority of the poor were the 'working poor'. They tended to work in sectors which were becoming structurally redundant to the national economy, and which were unproductive and unremunerative. This situation applied to male workers in particular, as most of the jobs created in the manufacturing sector during the 1960s and 1970s were for unskilled female labour (Cheah, 1977: 62).

Retraining facilities were grossly inadequate, providing facilities for training only one skilled worker for every eleven unskilled ones (Cheah, 1977: 69). Cheah (1977: 70) cautions:

The narrowing of opportunities for the poor, resulting from the displacement of the working poor from various jobs which they have previously engaged in, the rise in the general level of technical skills demanded, the increasing demand for formal qualifications, and the absence of adequate training and retraining opportunities, all serve, in the long run, to threaten their total elimination from the economy.

Since Cheah's work was completed, there has been an expansion of training facilities for school-leavers and a new adult education programme, but these still reach only a fraction of the unskilled labour force.

The reasons Cheah gives for the increase in poverty are deeply rooted in Singapore's economic system. It is the philosophy of the Singapore government's economic planning that rapid growth must exist prior to equality of distribution of rewards (Cheah, 1977: 13). High wage differentials which provide an incentive to effort and training are considered functional for economic growth and these differentials have been positively fostered in the National Wage Council recommendations which are a key part of the government's economic programme (Cheah, 1977: 75; *FEER*, 1.8.80: 46). It is also believed that subsidies of any kind reduce self-reliance, and subsidies on health care have been removed or reduced (*FEER*, 15.9.83: 32). The compulsory savings programme (CPF) and the rehousing programme which are measures aimed at ensuring the welfare of the elderly and the adequate housing of the population, have both had the short-term effect of burdening the poor, whose immediate budget for food, transport, and education has been greatly reduced by compulsory savings and by higher housing costs. The majority of Singapore's population was rehoused from wooden village houses or shophouses into high-rise flats during the 1970s, making this a period of physical upheaval, and financial burden and adjustment, which affected the low-income group in particular. Since proportionately more Malays than Chinese were found in the lower-income group by the 1970s, the Malay position as a group relative to the Chinese, declined further as a result of these trends and policies that tended to increase the disadvantages experienced by those who were already economically vulnerable.

The government's view of economic differentiation is that economic growth will eventually benefit everyone in Singapore, and that an open and accessible system of education ensures that no ethnic group or economic stratum of the population, nor any individual, is systematically excluded from obtaining a share of the highest rewards. Since educational opportunities are so important to the legitimization of economic inequalities, it is necessary to examine the extent to which equality of opportunity through education is really sought or accomplished in Singapore.

The National Education System

Although the principle of equal opportunity through education is given much importance in most industrial societies, whether capitalist or socialist, free mobility through education has proved to be impossible to achieve in practice. The material and educational resources of a student's home environment translate almost inevitably into advantages at school, whether through the direct role of finance for books, fees, and tuition or indirectly through the expectations of the student and his or her parents and teachers about the student's ability and potential future. Very few children of professionals end up as manual workers, and although a few individuals from lower-income backgrounds do rise through the education system, the majority do not.[4]

The Singapore government recognizes the connection between home background and educational performance, and has on various occasions provided statistical data to prove that the children of professionals do better in school than the children of manual workers. The government does not give priority to achieving equality of performance from students of all backgrounds, since that is regarded as impossible. It considers that Singapore has instituted a meritocratic system of education by ensuring that the best educational facilities are open *in principle* to all Singaporeans.[5]

In the 1970s and 1980s, the government allocated vast resources to education, with the aim of raising the overall educational level of the population as a key part of the economic development strategy to upgrade the work-force for the high-technology era. There had been severe problems of educational wastage during the 1960s and 1970s which resulted in more than 60 per cent of school-leavers entering the work-force with only primary education, functionally illiterate, and unable to use English, which is the principal language of government and industry. A new education system was introduced in 1979 to rectify this problem by a system of streaming at primary and secondary levels. Although again it was shown that students from better educated, more affluent homes were consistently streamed to the higher levels while the rest were channelled to technical streams, this inegalitarian trend was thought to be inevitable. The government defended the system by claiming that streaming ensured that poorer students would remain in school, and obtain a more efficient and extended education than they had obtained previously.[6]

The belief in the inevitability of the fact that educated and successful parents will tend to produce the nation's intellectual élite in a cycle that continues from one generation to the next is carried a good deal further in Singapore than in most democratic countries. The Prime Minister of Singapore has made public his conviction that intelligence is predominantly determined by genes rather than by environment, and he has instituted a number of programmes to ensure that the presumed genetic potential of the successful is nurtured while the reproduction of the unsuccessful is restricted. The primary purpose of the Eugenics

Board, established in 1969, was to restrict the birth rate to two children per family in order to reverse the tendency of the poor to have more children than the rich which would leave society with a large number of the 'physically, intellectually and culturally anemic' (Lee Kuan Yew, quoted in Chee and Chan, 1984: 7). In 1983 the Prime Minister expressed concern over the low reproduction rate among the nation's graduate women, and declared that this would diminish the nation's pool of talent with devastating effects on Singapore's future political leadership and economic survival (Lee, 1983). The government launched programmes to encourage educated women to marry and reproduce, a key incentive being a guarantee that their children would obtain a privileged education in one of the élite schools. A few months later another policy was introduced, offering $10,000 to couples with low education and low incomes who agreed to be sterilized after having one or two children. Although this was presented by ministers as a purely economic measure to help the poor to break out of the cycle of poverty by restricting their family size, and giving them cash to help them own a home, its eugenic implications were not lost on the Singapore population, and this and the pro-graduate mothers policy were later withdrawn as they proved to be politically costly.[7] The Singapore public was presented on numerous occasions with statistics on the above-average performance of the children of graduates which were apparently intended to prove the inheritability of intelligence.[8] This same data could also support the pro-nurture view that it is the whole range of social, economic, and educational advantages enjoyed by the children of graduates which enhance their chances of success.

Whether the high probability of success among the children of the successful, and failure among the children of the unsuccessful, is thought to stem from genetic or environmental factors, there are various features of the education system in Singapore which help to ensure the fulfilment of this prophecy. The most important factor lies outside the direct control of the government, and is the fact that only a minority of Singapore students speak the school language, English, in their homes. This minority is the offspring of those people educated in English for the previous one, two, or even three generations, who form the major part of Singapore's professional élite. In 1980, only 9 per cent of Chinese households and 2 per cent of Malay households spoke English as the principal home language (Census 1980: VIII.106), although there are other households in which the parents can speak English and endeavour to do so in order to improve their children's performance at school. Throughout their school careers the English-speaking students have an advantage over those whose first exposure to English is in primary school, and who take years to obtain the language competence of their peers, by which time they have been irrevocably branded as non-academic and channelled to vocational streams.[9]

Partly because of the issue of linguistic competence, teachers and principals are known to allocate students to graded streams before their first day in primary school, based on the student's 'biodata', which comprises information on the parents' education, occupation, and language accomplishments (*FEER*, 6.8.76: 46). Although this practice has been discouraged, teachers reported that it still continues in some schools. Educationists, academics, Members of Parliament, and the public expressed grave concern that the 1979 streaming policy would formalize and entrench the tendency of teachers to expect poor performance from students of uneducated home backgrounds, who would be most unlikely to have shown their full potential by the age of nine, which is when streaming takes place.[10]

The system of élite schools is a further factor which decreases the chances of upward or downward mobility through education in Singapore, and helps to maintain the present distribution of advantages and disadvantages. The government has acknowledged that there is a wide discrepancy in standards between schools, and recognizes that this is due to the better quality teachers, and higher percentage of students from better home backgrounds in the mission schools and a few of the best government schools (Goh, 1978: 3-11). The most important criterion for obtaining admission to one of the mission schools is for the prospective student to have an existing connection with the school: siblings in the same school, parents educated in that school or on its various committees, or with some connection to an affiliated school (*ST*, 29.7.79). This system ensures that families who had access to privileged mission education in the past will pass it on to their children. The criterion for admission to the best government secondary schools is 'merit' or performance in the primary school-leaving examination, and this systematizes the differentiation of government schools into 'good' or 'bad', since all the children with poor results are grouped in schools with a poor record of results, and tend to perform as predicted.[11] The government's concern to foster excellence in the good schools is not matched by a concerted effort, in practice, to ensure that every child has a real and effective, rather than nominal, opportunity to achieve his or her fullest potential through education.

Malays and Education

When educational statistics are compared on an ethnic basis, Malays consistently attain lower educational standards than Chinese. Table 7.13, compiled from census data, shows that in each age cohort more Chinese boys have proceeded to higher levels of education than their Malay counterparts (the pattern is similar for girls, who are excluded from this table for the sake of brevity).

There are three sets of factors that could account for the tendency of Chinese children to perform better in education than Malay children. The first factor relates to the correlation between home background and educational performance. The previous section indicated that there have been substantial impediments to the attainment of individual socio-economic mobility through education in Singapore in the 1959-80 period, and the tendency has been for the economic advantage or dis-

TABLE 7.13
Highest Qualification Attained by Malay and Chinese Males by Age Cohort, 1980

Highest Qualification	Age Cohort and Ethnic Group							
	15-19 Years		20-24 Years		25-29 Years		30-39 Years	
	Chinese	Malays	Chinese	Malays	Chinese	Malays	Chinese	Malays
No qualification (or only primary)	45.2	58.4	64.6	77.2	68.2	80.4	72.0	83.7
Secondary	8.9	10.5	18.6	18.6	15.8	15.8	12.9	11.6
Upper Secondary	2.9	0.7	11.7	3.1	9.8	3.2	9.2	3.8
Tertiary	–	–	1.0	0.1	5.8	0.5	5.9	0.8
Still in education	42.9	30.4	4.1	0.9	0.4	0.2	–	–
	100.0	100.0	100.0	100.0	100.0	100.0	100.0	100.0

Source: Census 1980: III.18-21.

advantage of a household to be sustained from one generation to the next. The economic and educational advantages enjoyed by the English-speaking, predominantly Chinese élite of the pre-war era have generally been passed on to their children, and there has been little mobility through education for the majority of Malays or Chinese, although the overall level of education has risen with the expansion of educational opportunities in the post-war years.

Since the economic position of Malay households relative to Chinese households weakened during the 1960s and 1970s, it is to be expected that this would come to be reflected in the educational performance of Malay and Chinese children. If the 1980 individual male income is taken as a very rough indicator of household welfare, 60 per cent of Malay households were in a weak economic position in 1980 compared to 40 per cent of Chinese (see Table 7.4). In the 1983 streaming exercise for entry to secondary school, 60 per cent of Malay students were channelled to the lower (normal) stream, compared to 40 per cent of Chinese students (*Key Educational Statistics*, 1983). The long-term consequences of this trend are disturbing, as the economic weakness of Malay households relative to Chinese, which arose during the 1960s and 1970s, is becoming consolidated through the educational performance of their children, and the economic gap between Malays and Chinese is likely to widen further.

Besides class factors, which have affected all Singaporeans (though Malays disproportionately, since more are poor), there are other factors unique to Malays which disadvantage them educationally in relation to Chinese. Up to 1965, about 50 per cent of Malay students were enrolling in education programmes taught in the Malay language (Lee, 1982b: 12). Although English education quickly took over in popularity after Singapore's independence from Malaysia in 1965, those students who began and continued their education in the Malay stream faced a range of problems. Until 1961, there was no provision for secondary education in the Malay language, and students who qualified for secondary schools had to transfer to the English stream, for which they were linguistically unprepared. Pre-university education was first made available in Malay in 1964, but there were never adequate numbers of trained teachers and textbooks at this level, a shortcoming reflected in the very poor results obtained by students studying in the Malay language at upper secondary level (Zahoor Ahmad, 1969; Sharom Ahmat and Wong, 1971; Gordon and Gunn, 1968-70).

For the few Malays who achieved the standard necessary to enter university, there were few courses taught in Malay, and Malay-stream students were effectively excluded from all the professional and scientific courses taught in English. Once they joined the work-force, they found that there were virtually no opportunities for Malay-stream graduates. Students educated in the Chinese language were similarly excluded from English-stream professional courses and jobs, but they had the alternative of a Chinese-language university and possibilities of employment in the Chinese-speaking sector of the economy which made their linguistic background less of a handicap. All education in Singapore is now in the English language, but a large part of the present Malay work-force was educated in Malay, and suffers from inadequate educational and linguistic preparation for employment in Singapore.

Although by 1980 all Malays attended racially integrated, English schools, Malay students still suffer from some disadvantages related to ethnicity. Within schools, it is common for Malays to be grouped together in one class for all subjects on the basis that this eases time-tabling for their second language (Malay). Given the tendency to prejudice and stereotyping which has been noted, many Malay students, parents, and leaders have found that classes in which Malays predominate are treated differently from those comprising Chinese and other ethnic groups. For example, a Malay boy who achieved the top grade in science in his class was forced to take technical studies instead of science because the timetable was not arranged to allow students taking Malay to also take science. This is not a deliberate government policy to disadvantage Malays but the result of carelessness or individual acts of discrimination by some teachers and school principals, and there is no way of telling how widespread or how significant this factor is in the Singapore education system.

In addition to the economic and ethnic factors that affect Malay educational performance, it remains to consider whether the poor educational performance of Malays when compared to Chinese on an ethnic basis, stems from factors within Malay culture. There is a strong conviction among some members of the Malay élite, the national government, and among many ordinary Chinese, that Malays are less hard-working and less achievement-oriented in education and in the economy generally than the Chinese, and that inappropriate cultural values account for their poor educational and economic performance. The origins of this view, and the reasons for its prominence, are considered in detail in Part III of this book. The description of Chinese household dynamics in Chapter 10 will show that there are some important differences between the cultural form of Malay and Chinese households, particularly in the relationship between the generations, which have some effect on their educational performance. It was argued earlier that the economic disjunction between the generations in the Malay household weakens the incentive to invest in children's education, while it will be shown that Chinese have a greater tendency to pool household incomes and to invest in the education of one or more family members as a means of gaining status and economic benefit for the parents and the family as a whole.

Although there is a difference in the householding pattern of Malays and Chinese that gives the Chinese an advantage in achieving mobility through education, the importance of this should not be exaggerated for two reasons. First, the comparison of Malay and Chinese educational performance given in Table 7.13 shows that, while 77 per cent of Malays aged 20–25 failed to gain more than a primary education, 65 per cent of Chinese in that age cohort were in the same position, so there do not appear to be strong grounds for the view that Malay culture is uniquely unsuited for educational achievement while Chinese culture is ideally suited. There are differences in the degree of emphasis that various cultural traditions give to education, but no absolute difference in kind, as all parents provide for their children's education to some degree. The second reason why the cultural differences between Malays and Chinese should not be over-stressed is that the correlation of home background with educational performance applies to both Malays and Chinese. In the case of both Malays and Chinese, the socio-economic status and economic conditions of a family influence practices of daily life and attitudes towards education. It was mentioned earlier that higher-income Malay parents pay more attention to their children's education than do low-income Malays, and the following chapter describes some of the strategies by which Malays in a favourable economic position attempt to ensure that their advantage is passed on to their children. Since practices and strategies towards education differ between higher- and lower-income Malays, it cannot be argued that there is a unique or unchanging aspect of Malay culture which forms a permanent impediment to educational achievement. The cultural idioms and some of the strategies by which Malays and Chinese attempt to achieve mobility or preserve class status do differ, and some of these differences may affect their relative educational performance to a degree, but, in general, children in Chinese households in the lower-income group experience many of the same problems as their Malay counterparts: an inability to cope in English, low aspirations, low morale, consistent failure in exams, and the need to augment personal or family income by beginning work early (see Chapter 10).

While it has been indicated that there are reasons to be cautious when analysing the impact of Malay and Chinese culture on educational performance, the statistical material presented in this chapter has shown that there are two areas with significant economic impact in which the practices of Chinese and Malay households differ: the extent of participation in entrepreneurship and the labour-force participation of women. These differences have contributed to the decline in the Malay position in the national economy relative to the Chinese. As the Malays have become concentrated in the lower-income group, their relative position as an ethnic group has declined further due to competitive disadvantages that affect everyone in the lower levels of the national social and economic hierarchy. The following chapters analyse in more detail the practices of daily life which, through their cumulative effects, structure the national trends towards class and ethnic differentiation described here. The chapters focus on the cultural basis of class differentiation in the Malay community, on Malay entrepreneurship, and on differences between Malay and Chinese households.

THE NATIONAL ECONOMIC AND EDUCATION SYSTEM 121

1. According to a 1983 newspaper feature, $590 per year is needed for the school expenses of a primary school pupil and $982 a year for a secondary school pupil, excluding the additional cost of private tuition (*ST*, 2.8.83).

2. A series of letters to the *Far Eastern Economic Review* in 1983 and 1984 commented on various types of discrimination against Malays believed to emanate from the government, such as the weak representation of Malays at the higher levels of politics and administration. See *FEER* (20.10.83; 8.12.83; 16.2.84; 26.4.84; 26.7.84; 16.8.84). The government maintains that recruitment to all government posts, including the armed forces, is based on merit not ethnicity, and the lack of Malay representation only reflects the shortage of qualified Malay candidates. See Ismail Kassim (1974) for an account of the reaction of Malay organizations to the National Service issue.

3. See Stough (1983: 237) for an example of sterotypes held by a Chinese factory foreman in Malaysia, concerning his Malay workers.

4. See Willis (1977) for a full discussion of the factors which tend to perpetuate class differentiation despite the apparent opportunities provided through education.

5. See *ST* (1.9.79; 11.9.79; 13.3.84) and Goh (1980) for comments by the Deputy Prime Minister on the question of equality, class, and mobility in Singapore. See *ST* (23.8.81 and 18.8.82) for statements by the government on the importance of raising the educational standards of the masses.

6. See *ST* (4.7.82) for statistics on the correlation of streaming with home background, which showed that 19.2 per cent of the children of skilled and semi-skilled workers were channelled to the lower streams compared with 7.2 per cent of the children of professionals. For a defence of education policies, see *Parliamentary Debates* (16.3.83: 977, Tay Eng Soon); *ST* (6.3.81 and 29.4.81 (letters, Goh Kim Leong)); *ST* (2.12.82; 28.8.83; 6.11.83).

7. For an account of these policies and their reception, see *FEER* (8.8.83; 6.10.83; 3.6.84; 21.6.84). For a full discussion of genetic fallacies, 'meritocracy' in education, and the genetic beliefs and policies of Lee Kuan Yew and the Malaysian Prime Minister, Mahathir bin Mohamed, see Chee and Chan (1984). Many believe there is a covert racial element of Lee's genetic views, and Mahathir has been explicit on this point. Eysenck was cited in the Singapore press as having found that in Malaysia, Chinese have IQs about 15 points higher than that of Malays, *Sunday Monitor* (21.8.83).

8. This data is given in *ST* (22.1.84; 20.9.84; 11.10.84).

9. See Gopinathan (1979) for a discussion of élitism in Singapore's education system and its relation to language competence.

10. For concerns within Parliament on the New Education System, see *Parliamentary Debates* (28.3.79: 101, Fong Sip Chee; 30.3.79: 254, Augustine Tan; 30.3.79: 275, Sia Kah Hui; 16.3.83, Sah Koon Seong and Tan Cheng Cock). In the press, see letters in *ST* (11.2.81, Leong Choon Cheong; 4.4.81, Pang Eng Fong and Linda Lim; 4.4.81, I. C. Johnstone; 17.8.82, 'Conjecture') and see also the earlier discussion in *Commentary*, Pang and Lim (1976).

11. See Mansor Haji Sukeimi (1979) for comments on inequalities between schools. It appears that not all parents are aware of the great discrepancies in standards between schools, or else they are obliged by their low incomes to reduce transport costs and send their children to the nearest school regardless of standards (*ST* (21.8.79)).

8 The Cultural Basis of Class Differentiation

THE preceding chapter described the structure of educational and work opportunities within which Malay and Chinese households strive to sustain themselves or to improve their lot. It established that there are significant income differentials between sectors of the Singapore population and argued, with special reference to the education system, that there is a tendency for the emergence of social classes that are perpetuated from one generation to the next. Although the opportunities of an individual are structured by ethnic factors and by the economic position of the household into which the individual is born, they are not entirely determined by economic, social, or institutional conditions. It is a characteristic of democratic capitalist societies that there are significant areas of choice for the individual, such as whether or not to study hard in school or to start a business. Individuals make decisions in the light of their cultural knowledge and it is through these culturally informed practices that individuals come to be channelled to higher or lower positions in the national socio-economic hierarchy. In turn, a cumulative and unintended outcome of these culturally informed practices is to sustain and institutionalize the hierarchical conditions within which individuals and groups find themselves constrained.

This chapter explores the impact of economic factors, especially incomes, on the practices of Malay households and considers how Malay cultural practices in turn sustain class differentiation. The terms 'higher-' and 'lower-income' are used here as a convenient reference to the economic hierarchy, but the picture is complicated by the presence of households with more than one income-earner, and by variation in the proportion of individual earnings that are pooled in the household. Here incomes, education, and other economic factors are not treated as unilateral determinants of cultural practices; instead the investigation focuses on the ways in which class is achieved and displayed in the culturally meaningful practices of daily life. The set of concepts and practices related to social hierarchy in contemporary Singapore Malay society is a unique cultural configuration, and the chapter examines its relationship to material conditions, to ideas deriving from the larger Malay cultural heritage and Islam, and to other features of the national economic and cultural environment, especially the presence of the Chinese.

The Impact of External Conditions on Household Form

The structure of Malay household relationships in contemporary Singapore described in the first part of this work does not differ fundamentally between households of different economic status. This can be attributed both to the deeply rooted nature of the individualistic pattern of householding relations in Malay culture, and to the relatively narrow range of economic conditions encountered by Malay households in Singapore. The previous chapter noted that most Malays are concentrated in the lower-income group, and the range between the highest- and lowest-income earners is not extreme, as there are few very wealthy Malays, and none who are completely destitute or homeless.

A common source of difference between higher- and lower-income households of the older generation in Singapore was the pattern of fertility, which was related to the need of the poor to have more children for old-age security. Today, the Singapore government's compulsory saving scheme ensures that even low-wage earners will realize a substantial capital sum on their retirement or will own a government flat, and this means that, like their higher-income counterparts, they will not be totally reliant on their children in old age. This has probably been a significant factor in the drastic drop in the birth rate of both Malay and Chinese families. The smaller family size which characterizes the younger-generation families in Singapore of all income groups, is accompanied by more intense relations within the nuclear unit, and more focus on children and their activities, than is found in the older generation. Small family size and intense focus on relations within the household are features commonly associated in the West with the middle or upper class, but in Singapore this pattern now characterizes most of the younger-generation households.

The rehousing of the population in government-built flats has further contributed to the intensification of relations within the nuclear unit for all income groups. Where previously, lower-income families living in squatter settlements or crowded shophouses had little privacy and were often obliged to share water and other facilities with neighbours, now, even they can close the door of their flat and sustain intense interaction as a family within their own private space if they so desire. Although lower-income families in small flats do have more uncontrolled interaction with neighbours than higher-income families in large flats or houses, the difference in the degree of privacy they can enjoy is no longer so great.

Another factor which affects all Singapore households is the bureaucratization of the housing, education, and employment systems, which again has served to intensify relations within the household. Lower-income households in Singapore do not generally find it necessary or useful to cultivate relations with patrons or with kin and neighbours in order to gain access to work, housing, or other scarce resources. Contacts can be of some help in providing recommendations for jobs, but this was not crucial during the period of research when Singapore enjoyed full employment. In general, Singapore households can only progress and obtain access to these resources through their own efforts and in direct dealings with the relevant government departments. As the need for reliance on others has been reduced, the need for the household to conserve its resources for its own consumption needs, and for the education of children, has risen and this has been a factor shaping relationships inside and outside the household for all income groups.

The factors in the economic environment of Singapore households discussed here–pension provisions, restrictions on family size, rehousing and bureaucratization–were all planned by the government as measures for the welfare of Singapore, although the government probably did not predict, or intend, the effects these measures would have on the form of family life. These factors have made the economic conditions faced by the majority of Singapore Malay households relatively uniform. To understand how differences in status between households are achieved, sustained, and displayed, it is necessary to look at some of the more subtle differences in the practices that characterize higher- and lower-income groups, especially the strategies adopted in dealing with individuals outside the household.

The Sense of Class and the Achievement of Social Mobility

Malay society places considerable emphasis on the relationship between the individual or household and the social circle of kin, neighbours, and friends. While wealth is something an individual Malay in Singapore can gain through his or her participation in the national economy, to enjoy status in the Malay community it is necessary to sustain a Malay social circle, since status is ultimately something *granted* by others. This can be a source of difficulty and ambivalence, since attaining social mobility involves a process of detaching oneself or one's household from a current circle of peers, and gaining acceptance in another circle deemed by some set of criteria to be better or higher. A close examination of the type of relations an individual or household has, or claims to have, with the social circle, gives an indication of the way social status is perceived, gained, and demonstrated in Malay society.

When discussing social relationships, Malays of any age, sex, or income group often stress that they have little interaction with people outside the household, and they prefer to stay home unless there is an urgent or important requirement to go out. The frequency of this claim has been noted by other researchers on Malay society (Normala Manap, 1983: 111). The essence of the claim is that, by keeping to oneself, one does not become involved in improper scenes such as gossiping, borrowing and not returning things, children's quarrels, prying into affairs of others by free visiting in their homes, or jealousies over property which sometimes result in accusations of witchcraft. All these are commonly acknowledged to be coarse, non-refined forms of behaviour, associated with people of low education and low socio-economic status. Claiming to stay at home is not only an assertion that one is *apart* from these undesirable neighbourhood scenes, but also a subtle assertion that one is *above* them.

Keeping a distance from Malay neighbours and peers is a claim which indicates a perception of class. It is also a strategy by which economic mobility can be attained. Parents who wish to keep their children in school try to restrict their interaction with peers who may not share their aspirations. Where boys have fallen into trouble, grown long hair, or dropped out of school, this is always blamed on their mixing with bad sorts, in particular their peers in the neighbourhood or the Malay community at large. The influence of peers in causing children to leave school early is not unique to Malays or to Singapore. Chinese, too, in the life stories collected by Leong Choon Cheong (1978) describe how, seeing their siblings and friends earning money having left school early, they wanted to follow suit. Neither is the influence of peers the only factor in early school-leaving, since children from less-educated homes are likely to be already alienated from school, having failed exams repeatedly. But restricting the interaction of school-going children with working or non-studious children is one essential part of a strategy of upward mobility. It is a strategy Malay parents have been advised to follow in various programmes to improve Malay educational performance, which will be described later.

The difference between lower- and higher-income households lies not so much in their assessment of the dangers of mixing freely with others, but in their effectiveness in restricting interaction. Though some lower-income households make the *claim* to stay indoors, their housing is cramped and provides less opportunities for securing privacy and keeping youngsters indoors, away from neighbours. In most lower-income households, both parents are working or the father is working at two jobs, reducing the intensity of parental control and supervision of children. In higher-income families, too, the wives often work, but the preferred pattern is for the wife to stop work once the children begin school, precisely in order to supervise their studies *and* their social interaction. If the mother continues to work, higher-income parents plan their child care provisions on the basis of the educational standards and interaction patterns of the child-minder. They often prefer to place a child in the household of an educated non-working sister or neighbour of a similar socio-economic and educational standing, rather than with an uneducated mother, especially if the mother tends to be 'too busy with her friends'. Lower-income families are less likely to be able to afford any form of child care when the children reach school age, and tend to leave them alone or with neighbours where their interaction is uncontrolled.

If the avoidance or careful control of social interaction with the Malay community is an important component of a strategy for social mobility, it also has a social cost. Those neighbours or kin to be avoided are those of lower or equivalent status whose level the social climber wishes to surpass. In restricting relations with these people, the individual or household risks becoming socially isolated. The term used to refer to people who keep to themselves is *sombong*, which means 'proud'. In each of the Malay neighbourhoods studied in the course of this research, there were individuals or households regarded as *sombong*, and though this is explained by neighbours primarily as an individual or personal characteristic, in each case the people involved were making a concerted and carefully planned attempt to achieve mobility, with or without a slightly higher than average income. Those with a higher income who keep their doors open and engage in free interaction with their Malay neighbours are the subject of positive comment–'he earns a lot but he is not proud'– and the implication is that this is rather unusual. Generally, it is assumed that once people are better off, they will forget their former circle of kin, neighbours and friends, and though expected, this is always a cause of bitterness. The cooling of relations is often instigated by the lower-status party, who withdraws in expectation of a rebuff.

The cost of this isolation from neighbours is not a financial one, since neighbours, kin, and friends give little direct financial help to each other in any income group. The cost is social. Although interaction with other Malays is fraught with risks and the dangers of gossip, social worthiness or goodness (to be *orang baik*) is only proven in relations with outsiders. The quality of being *orang baik* is an essential part of social esteem. It is not directly related to wealth, and wealth can actually be disadvantageous if the attainment, preservation, or enjoyment of that wealth leads to social isolation based on accusations of pride. But wealth is not irrelevant, since a characteristic of being *orang baik* is a reputation for refined behaviour and non-involvement in gossip and quarrels, which is easier to achieve from the privacy of middle- or higher-income housing.

Ceremonial occasions involving large numbers of people are a primary arena for the display of this quality of social worth. If a family suffers an illness or death, and kin and neighbours do not attend in large numbers to offer help and condolence, the explanation proferred by onlookers will be that the person was proud (*sombong*) or the opposite, a quarrelsome busybody, and so no one cared. It is important to be seen to have a good relationship with kin and neighbours. The number of people who come to help with wedding preparations, or who attend weddings, is a source of comment and pride. For example, on several occasions a wedding host made a claim that some former neighbours or friends, whom the host had not invited for fear of inconveniencing them as they lived far away, had nevertheless heard about the wedding and decided to attend. This was recounted with pride, because the guests' gratuitous presence, their willingness to undergo the trouble and expense of attending yet another wedding, was a sign of the high personal regard they held for the host. The number of 'adopted' children (or adults) who have voluntarily attached themselves to an older person is another sign that this person is good and worthy. When the adoptees visit the adopted parent, or offer help when the adopted parent is hosting a major event such as a wedding, their presence is a public statement about the social worth of the host.

The cultural configuration described here is one in which social status is achieved and proven in the relationships built up in personal interaction with others. This can be seen as a counterpart and continuation of the theme of individualism and personally created and sustained family relationships discussed earlier. In Malay society, the individual is expected to serve his or her self-interest, and yet the 'other' is an integral part of the individual's personal development and social- and self-esteem. It is a system of individualism, but not a system of independent atoms in which others are irrelevant. There is a very great stress on the process of sustaining relationships with others both within and outside the household, precisely because these relationships are not mandatory or taken for granted, and yet they are felt to be highly desirable.

Since all relations with kin or neighbours are, ultimately, optional and individually created, the voluntary presence of individuals at a ceremonial function or assisting in its organization is a social statement. It is best understood as a form of gratuitous gift, which through its voluntary and personalized nature conveys a message about the social worthiness and goodness of both visitor and host. The Singapore situation emphasizes the gift aspect of interpersonal relations, because kin are scattered and neighbours are *de facto* less close than in a stable village community, and most people have half a dozen or more wedding invitations for any Sunday at certain times of the year, besides alternative forms of entertainment and other demands on their time and cash. Their attendance at any particular wedding is thus a more powerful statement about the relationship between host and guest than in a village community where everyone would be invited and would probably attend. Consequently, attendance at weddings in Singapore is a greater source of anxiety, tension, and gossip. Wealth alone cannot prove this form of social status. Wealth can be used to pay for a very large wedding feast, but if people are not on good terms with the host, they may not attend, and will merely comment that the host is throwing money around and showing off.

It should be noted that a reputation for religiosity is not an alternative source of status, nor a proof of social worth (being *orang baik*). People who are 'always busy praying' or who show religiosity publicly by diligent attendance at the extra prayers during the fasting month are often suspected of insincerity. Religious people may be *orang baik*, or they may be *sombong* (though they are unlikely to be gossipy or quarrelsome), but whatever the case, their status and reputation are achieved through interaction with others and not in isolation.

Although individual religiosity is not in itself a source of status, the principal social occasions on which social interaction is enjoined and positively sanctioned have a religious element. Islam encourages communal prayers, and the most appropriate setting for interaction between men is thought to be in the nearest mosque or community prayer house, or a private house in which prayers are being offered for the dead. There are other occasions, such as religious classes and gatherings to pray for the safety of pilgrims, which are attended by men and women. Islam requires all Muslims to get to know their neighbours so that they can be informed of life-crisis events, such as funerals, illnesses, births, circumcisions, and weddings, which they have a religious and social obligation to attend. The religious aspect of the event provides a semi-sanctified environment for interaction, in which gossip and quarrels are less likely to occur than in informal contexts. The possibility of gossip is not excluded, however, and although diligence in attending these events can be a source of reputation for 'baikness', the voluntary presence (or absence) of an individual from any particular event is fraught with social implications. These events do not really provide neutral ground: they are arenas for the acquisition and demonstration of social status, and they must be carefully managed.

Public housing in Singapore groups households in blocks of flats on the basis of household income. This feature of urban design has made the immediate neighbourhood, which is the apartment block, relatively homogeneous in economic standing, and it has encouraged the emergence of distinctive forms of social life among the higher- and lower-income block residents. Once a household has acquired sufficient economic means to enable it to move into a high-income block, it is assured of gaining a new arena within which to interact with households of similar standing, and is able, physically, to leave its former peers behind. To some extent, this eases the tensions associated with the process of social mobility, although it increases the gulf between neighbourhoods and the feeling among the lower-income group that all those in the high-income blocks are *sombong* since they interact primarily among themselves in a way which tends to exclude others.

A distinguishing feature of the pattern of social life among Malays in higher-income apartment blocks is the restriction of interaction to formally organized, religiously sanctioned events and the diligence of households or the block as a whole in organizing these events. Often, Malay households form a block committee with a chairman, secretary, and treasurer and hold regular meetings and elections. They organize a range of events for block members such as religious classes, weekly Koran-reading sessions for men and women, youth groups to perform various services at weddings, and some annual religious events. Their most important role is to inform all the Muslims in the block of a neighbour's death, so that as many as possible can attend the gathering at the house of the deceased immediately. Usually a collection for the bereaved is made by the committee members or a standard amount is given as a donation out of the committee's funds. While this sum of money may be helpful or even crucial to help a lower-income family pay the immediate expenses of the funeral and the accompanying feasts, the provision of a few hundred dollars is not of great significance to a higher-income family; yet the block committees are at their most vigorous in the higher-income blocks. The reason for the prominence of these committees in higher-income blocks is that by organizing or participating in block activities, the members are able to show that they are good Muslims, and good, caring, and co-operative neighbours who are at hand should others need their help. They fulfil the Malay requirement for sociability, while avoiding informal interaction with neighbours which carries all the risks of gossip and quarrels which are indicative of low status.

There is virtually no informal visiting between households in high-income blocks, and every household is mindful of its privacy. There is also some nostalgia for the freer form of interaction of the kampong, when people could meet and chat on neutral ground outside their homes without a formally stated time, place, and purpose, or a prior warning telephone call which is the minimum requirement to arrange a meeting between adults or children in a higher-income block of flats. As one individual described life in a block of five-room flats: 'This block is like a kampong, we all know each other. The only difference is that we don't sit outside in the evening to chit chat. Here we get together only on certain occasions.' In saying this, he expressed his pride at the close sociability achieved among the Muslim flat-dwellers, which he likened to the kampong ideal. At the same time, he values the increased formality and control which characterizes life in a higher-income block. He described his former residence in a lower-income neighbourhood as a slum, where children were not controlled by their parents and did not study. It was to ensure his children's education, either by keeping them home or by allowing them to associate only with suitable companions, that he chose to move to his present neighbourhood.

In lower-income blocks, the same religious events are considered to be important, but the organization is likely to be less formal because the households, especially the women, are already in daily, informal contact with each other, with all the attendant risks. The factions and quarrels which result from close proximity in a lower-income block are often made clear in the pattern of participation at religious feasts or life-cycle events held by individual households and sometimes these factions prevent the formation of an effective block committee. One lady made the observation that 'in purchased flats, the important thing is religion; in rented flats, it's only fighting'. Although exaggerated, this comment does capture the essential difference in the way interaction among higher- and lower-income groups is organized and perceived.

Individual and Community: The Cultural Heritage

It is useful to examine in more detail the sources of the hostility and accusations of pride that are levelled at individuals or households who are engaged in the process of social mobility. It was observed in Singapore that relations between two parties who sense themselves to be of unequal status, whether by economic or less explicit social criteria, are almost invariably tense and uncomfortable. Such is the case whether the parties are close kin, such as siblings, distant kin, former friends, or neighbours. A possible explanation for awkwardness or even hostility between parties of unequal status is that Malay society, in common with many small-scale and peasant societies, may have an egalitarian ethic which requires that wealth or poverty should be shared, and that no individuals should seek to raise themselves above the level of the majority.

In other cultures where this ethic is reported to be present, conformity is enforced through social pressures such as gossip, isolation, and accusations of witchcraft. These are the sorts of pressures which are experienced in the Singapore Malay community, and the question to be addressed here is whether or not these tensions stem from an egalitarian ethic.

In Singapore, Malays are seldom involved in direct economic relations with each other as employer and employee. The material gain of one Malay party does not result in a direct material loss to another Malay, since all Malays gain their livelihood through their participation in the national economy. There is virtually no economic aid between Malay kin or neighbours in Singapore, and the poor do not expect any aid to be provided as a matter of course. Instead, they claim that it is 'everyone for himself', and even the most economically vulnerable households, such as those of widows, claim that they have not been offered any material help by their kin, nor did they feel they would be likely to receive any aid if they did make a request.

The following is an example of kin assistance in an emergency situation: A low-income family earning $500 per month suddenly needed to raise $2,400 for the hospital expenses of the wife in delivering her third child. A sum of $900 was obtained from the husband's Post Office savings; $600 by pawning all the wife's gold; $300 from the Post Office accounts of the two children (to be repaid to them); $200 from the wife's father, which was seen as part of an exchange because she had sent him $200 earlier when he had suffered a fire; $200 from the wife's sister, which she plans to return by giving $50 on the occasion of each of the sister's children's weddings; and $200 from the wife's stepbrother, who was considered by the wife to be unusually close and supportive. The husband's first thought when he learnt of the enormous sum required had been to ask for assistance from his sister, who had received more than a million dollars as compensation for some land. 'I took the bus and went to see her. I told her my wife has to pay $2,400; she said, "Oh, that is no problem, is it, because you can always pay by instalments," so I felt embarrassed to ask her. She never asked me whether I had enough. Really, all my siblings should have helped me with $100–200, if not given then at least lent, but I was afraid they would just say "Why haven't you saved any money?", so I did not ask them, even though they knew about my difficulties.'

The observed absence of aid between Malay households in Singapore could imply that although there is envy of the wealth of others, this is not related to a view that wealth should be shared. Alternatively, it could be that the belief that wealth should be shared is present, having carried over from the rural cultural heritage, although the practice of sharing has broken down in Singapore, resulting in tensions arising from unmet expectations. On this point, it is useful to turn to the rural ethnographic literature.

Several writers on rural Malaysia and Java have argued that there is a form of communitarianism in the village economy. Clifford Geertz (1956: 141) writes about shared poverty among the Javanese *abang-*

an and describes the relationship as one based on a commitment to 'a communalistic rather than an individualistic approach to economic problems', in which food or its absence is shared equally among the families in a village. Scott writes of 'communitarianism' in Malaysia and 'village redistributive pressures' which ensure a minimum subsistence for all the inhabitants. He relates this to social pressures such as jealousy and envy which discourage the desire to outdistance fellow villagers and ensure that any accumulation will be redistributed. Scott (1972: 15) mentions Foster's discussion of peasant society and the Image of the Limited Good. Swift (1967: 261), writing about Malaysia, also notes that there are pressures to secure redistribution through the requests of the less prosperous for loans that are not repaid; the pressures are reinforced by mockery, criticism, and hostility towards the rich. Like Scott, he considers (1965: 154) these pressures to have the material effect of limiting the incentive of rural Malays to acquire wealth.

The analysis of South-East Asian society in terms of shared poverty and the redistributive ethic has been challenged on several grounds (Alexander and Alexander, 1982; Diana Wong, 1983; Stoler, 1977; Hart, 1986). One criticism important to the present argument is that the village, which is the supposed framework of communitarianism, does not act as a unit in the way the accounts of Clifford Geertz and Scott suggest. Diana Wong found in her village study in Malaysia that acts of charity, such as giving rice to the elderly poor, are individual acts seen in the context of Islamic obligations and religious merit. She reports that the poor who are not elderly, generally go hungry or migrate out of the village. There is no corporate village or individual action taken to provide for them (Diana Wong, 1983: 283, 297). Hardjono (1987: 203-4), writing on West Java, has similar observations.

An important relationship in the rural economy is that between landlord and tenant or share-cropper, since it is concern for the fate of the weaker party in this arrangement that has a direct bearing on whether the poor have enough to eat in a year of poor harvest. There is some evidence that landlords do tend to be flexible in this regard (Bailey, 1983: 39), and it may well be significant in restricting the accumulation of wealth by some households, but three important points must be noted. First, this is a relationship between two individuals and does not involve the village as an entity. Secondly, close attention should be paid to the kinship relations between the parties, since this may be the factor that puts pressure on the landlord to treat the tenant with consideration (Bailey, 1983: 41). Insufficient attention has been paid to the specifics of kin relations, perhaps because of the common Malay claim that all the villagers are related or act 'as if' they were related. Diana Wong found in Malaysia that flexibility in exacting rents or shares, or a land-owner's willingness to rent land to a tenant when the owner could have profitably farmed it himself, are most often found between parents and children or adoptive children, and should be seen as part of the process of the devolution of property between the generations. In general, the rents or shares expected from kin and non-kin are the same, although kin may be given preference over non-kin in securing a tenancy when one becomes available (Diana Wong, 1983: 282, 285; for Java see Hardjono, 1987: 66). Thirdly, where generosity is shown, whether between parent and child or any other parties, it is seen as such: a personal act of generosity in which an individual chooses to forgo his right to fully enjoy his or her own property out of love or sympathy for another. This aspect of interpersonal relations was discussed at length in the earlier account of the Malay family. Giving or sharing stems from the generosity of the individual will, and does not stem from any view of the land or its products as being corporate family *or* village property. 'Communitarianism' is an inappropriate term to use for this set of views about people and property.

There are other practices in Malay and Javanese villages which have convinced some writers of the existence of a communalistic orientation, and these are known by the term *gotong-royong*. A study in Java by Koentjaraningrat (1961) analysed the various systems under which villagers labour together on projects for the benefit of the whole village or projects that benefit a particular household. Although villagers may help a household with agricultural work, house-building, or by making cash or labour contributions to the hosts of a major feast such as a wedding, Koentjaraningrat found that balanced reciprocity is the rule. A recipient of aid incurs a debt to return the aid in like form at a later date, and sanctions such as gossip and hostility are applied to those who do not meet their obligations. The only time that uncalculated aid is given by villagers is on the occasion of bereavement which, as Koentjaraningrat notes, is in almost every culture an occasion on which help is offered freely. Accounts from rural Malaysia and Indonesia such as those of Bailey (1983), Diana Wong (1983), and Jay (1969) describe a similar set of arrangements for mutual aid, and argue that it is not of major importance to the economic welfare of village households.

Contrary to the assertion that poverty is shared through mechanisms for redistribution, the ethnographies noted above indicate that in rural areas of Malaysia and Java, the household acts as a fully independent unit, which relies on its own resources for its daily sustenance and, also, in the long term, for its special or ceremonial needs. In urban Malaysia, Nagata's (1976) account of the relations between kin concurs with the situation described in Singapore, and even McKinley (1975), who has made much of the importance of 'collateral kin security' in urban Malay society, acknowledges that very little material aid actually passes between kin.

Observations in Singapore, and comparisons with other times and places in the Malay world, suggest that the tensions that characterize the relations between parties who are socially and economically unequal in Singapore Malay society, derives not from failed communitarianism, but from the ideas of individualism and the counter-balancing stress on building and sustaining relations with others, which were discussed at length in the analysis of the Malay family in Part I, and mentioned in the description of Malay community life in Singapore earlier in this chapter.

It is expected that the individual can and will pursue personal economic goals. But the individual is not a recluse, and social relations with others both within and outside the household are highly valued in themselves while also being fundamental to social- and self-esteem. In order to build and sustain relations with others, it is necessary to treat them as whole social beings, with sympathy and consideration. Sustaining a relationship involves 'giving' a little of the self, which is often seen as related to a willingness to give up material values, or make gifts. It is the desire to maintain good relations with others that places a restraint on the overt, unmitigated expression of calculative self-interest. The individual who drives a hard bargain or who insists on accumulating personal wealth without regard for the needs of others is disliked not because he or she is upsetting a communitarian ethic, but because that individual is upsetting, or failing to build and sustain, social relations with others. Failure to aid a tenant, neighbour, or kinsman who is without the means of subsistence in the rural economy, or failure to aid urban neighbours or kin, does not result in anger directed at the material transaction, since the owner is recognized to have every right to his or her property. Instead, it results in deep personal bitterness and hurt, from which all parties lose in honour and esteem. Both the more and the less affluent Malays in a neighbourhood feel permanently anxious and become extra sensitive to possible snubs, vulnerable to isolation, precisely because they *want* to be on good terms with others.

The importance given to interpersonal relationships, and the tendency to see all relationships in terms of an inseparable combination of social and economic elements, makes face-to-face direct economic dealings between Malays tense and problematic. In the Malay view, there is no 'pure' personal or 'pure' economic transaction. Swift (1965: 171) writes that rural Malays 'continually weigh their economic relations with kin against a purely economic relation with a non relative, and express a clear preference for the latter'. The requirement to regard the other as a whole social being increases with the closeness of the social bond, although there is no radical disjunction between the behaviour towards kin and non-kin, as all relations with others tend to become personalized. In Singapore, where Malays earn their livelihood in a non-Malay context, they are not hindered or withheld from economic advancement by this set of views about social relationships, and they manage any stress or tensions that arise from their relative affluence by the strategies described earlier. The cultural configuration described here does place some restraints on the pursuit of wealth where it involves direct economic dealings with other Malays as in the case of entrepreneurship, the subject to be discussed in detail in the next chapter.

Class and Hierarchy in the Presence of the Chinese

It was noted earlier that the participation of Singapore Malays in the multi-ethnic national economy has affected the pattern of Malay social differentiation, because it has provided a socially neutral arena for economic gain in which Malays are economically independent of each other. It was also indicated that wealth and education, which are criteria of status in the national hierarchy, are integrated as part of the Malay view of status, although they do not wholly define it. There is a further way in which the presence of the Chinese has affected the patterns of social life that have developed among higher- and lower-income Malay households, and that is in the tendency for social status to be associated with, or even rationalized by, cultural differences between ethnic groups. The tendency is related to the process of cultural involution or the ethnic group's search for signs of cultural distinctiveness.

Malays often attribute the problems of gossip, quarrels, and tensions in interpersonal relations, to a special Malay failing, and contrast this failing with the behaviour of the Chinese, who are said to be less sensitive, less nosy, and who like to keep to themselves. These common observations on the differences between the behaviour of Malays and Chinese as neighbours result in the interesting situation where many Malays, especially the upwardly mobile who are most vulnerable to gossip and anxiety, state a preference for Chinese neighbours because they consider that the Chinese will not intrude or interfere in their personal or family lives. While every Malay neighbour is a socially relevant person who has to be greeted and acknowledged if sensitivities are not to be offended, Chinese comprise a whole category of people who can safely be placed at a social distance on the basis of ethnicity, and can be ignored with no repercussions. The Chinese have taken on the curious role of 'spacers', who put a physical distance between one Malay family and another and help reduce tensions among Malays, while to Malays the Chinese themselves are socially irrelevant. From the viewpoint of non-interference, Chinese are ranked higher than Malays as good neighbours, mapping hierarchy on to ethnicity.

The Chinese are also regarded as 'better' than Malays in the field of education, and the entire ethnic group is credited by many Malays with the qualities of diligence and seriousness in studies. The efforts that Malay parents make to control their children, especially boys, are directed at limiting their exposure to *Malay* peers. Malay youths' failings in terms of drug abuse and inattention to study are perceived by Malay parents to come from outside the home but *within* Malay society. This has led to a situation where many Malay parents or the students themselves choose schools where there are more Chinese and few Malays, in the hope that the student will learn good study habits from the Chinese while also avoiding Malay company. Some parents note with pride that their children have mostly Chinese friends, because they mix with fellow Chinese students at school and avoid Malay company in the neighbourhood. In the process of attempting to gain social mobility through education, it is not only non-studious children who must be avoided but *Malay* children in general, since non-studiousness is a trait taken by Malays (as well as non-Malays) to characterize Malays as an ethnic group.

The cultural characteristic of non-studiousness which is supposed to distinguish Malays from Chinese is only stated by Malays on a general basis, since individual Malays always have more specific accounts of the problems they faced in education, and they are also aware of non-studiousness among some Chinese. Nevertheless, the association of ethnicity with cultural traits that promote or impede educational achievement affects the way that stratification in the society as a whole is seen, and it affects the actions that individuals take. Many Malays do not perceive that the mission school students succeed because the schools are better or because the students come from homes which give them linguistic and other advantages. Instead, they consider it to be the Chineseness of these schools that breeds success, while it is thought to be the over-representation of Malays in the poor schools, or the influence of Malay peers, that causes failure in the Malay students who attend these schools.

On the other side of this interlinking of status and ethnicity, Chinese as a whole are ranked very low in terms of some Malay values. Keeping to oneself is a mixed virtue in Malay society, because while neighbours should not gossip and quarrel, they *should* be involved with each other to a degree. It is fundamental to the image Malays hold of themselves as an ethnic group, in the context of Singapore, that Malays are co-operative and caring neighbours in contrast to the Chinese who are said to only look after themselves. Close community relations are a focus of Malay ethnic pride, a pride reinforced by government campaigns to foster neighbourliness, since the Malays believe the campaigns to be directed primarily at the Chinese who fall short in this regard while Malays set the ideal standard. This is a further example of cultural involution, in which the image that Malays hold of themselves and of the Chinese as ethnic groups with distinct cultures has become incorporated into the practices of their daily lives. In part, the diligence of Malays in forming their block committees and the tacit obligation to greet all Malays encountered in the neighbourhood is related to the desire to prove and enhance their *Malayness* in the presence of the Chinese. Those aspects of behaviour that mark a Malay as a good and refined person (*orang baik*) also mark him or her as a Malay in the context of multi-ethnic Singapore and the idea of status is again linked to the ranking of ethnic groups.

Some practices within the household are seen in terms of ethnic contrast. Malays like to see their families as close, co-operative, and refined, in contrast to the Chinese, whom they sometimes overhear quarrelling and swearing inside the home, and who have been known to abandon their elderly in Homes for the Aged. Again, government campaigns encouraging filial piety are seen as directed at the Chinese, and they confirm the Malay sense of moral superiority in this regard. There are reported to be no Malays in Homes for the Aged (*ST*, 23.3.81). The opposite view is also held, that Chinese families are close and co-operative while Malays are 'each for his own': which idea is more prominent depends on the context.

When not making general, ethnic statements, Malays draw distinctions between different types of Chinese. Malaysian and Indonesian Chinese and the *Peranakan* (descended from early Chinese settlers) are often thought to be more courteous, having learned from the Malays, while Singapore Chinese, whether from Chinatown or the few agricultural settlements in Singapore, are known as 'hill Chinese' and felt to be the most coarse and 'low class'. Christian Chinese and those who speak English are regarded as refined through their association with British culture, and they are sometimes described as 'white Chinese'. They are thought to be less prone to discriminate against Malays. The system of ethnic stratification does not always predominate in the way Malays see Singapore society, and there are occasions when they feel affinity with their Chinese neighbours on the basis of being 'all poor people like us', and they react favourably to any Chinese who are friendly and open or, in Malay terms, not 'proud'.

The presence of the Chinese has had some profound effects on the shaping of Singapore Malay culture, especially through the phenomenon of cultural involution described above. Surprisingly, the extent of direct interaction between Malays and Chinese is rather limited. The public housing programme has ensured that every block of flats in every new town has a mixture of ethnic groups, but the privacy afforded by flat-dwelling has restricted the arenas for informal interaction between Malays and Chinese that existed in many multi-ethnic squatter settlements. In lower-income blocks there is some casual visiting, baby-sitting, food-exchange, and a familiarity based on proximity, especially between Malay and Chinese women, though the absence of a mutual language prevents deeper interaction. Higher-income Malays do speak English, and interact more with Chinese at work and in the course of higher education, but they have little informal contact with Chinese outside these two contexts. For these Malays, interaction with Malays in the neighbourhood and the focus on Islam have become the prime arena in which to preserve, demonstrate, and enhance their sense of Malayness *and* their sense of class, in the context of Chinese Singapore.

This chapter has described how social and economic features of Singapore, in particular the trend towards the formation of socio-economic classes and the presence of different ethnic groups, have influenced the cultural patterns of contemporary Singapore Malay community life. It is the practices of daily life in Malay households and in the Malay community that form the framework within which Malays attempt to gain socio-economic mobility, or demonstrate their class status. At the same time, it is through these practices, whether in direct or indirect ways, that Malay individuals become channelled to different positions in the national socio-economic hierarchy, and the uneven distribution of wealth and status is itself perpetuated.

9 Malay Entrepreneurship

THE low rate of participation of Malays in entrepreneurship has been an important factor shaping the position of the Malays in the Singapore economy. It accounts for the weakness of the Malay élite from early times in Singapore and it accounts in part for the weakening of the economic position of the Malays in relation to the Chinese, especially during the 1970s when business activities were a source of upward mobility for Chinese households. In a less direct way, the non-involvement of Malays in business has deeply affected the position of the Malays in the ethnic and cultural framework of the nation, because it has been an important source of the poor image non-Malays hold of Malay society. Almost one in four working Chinese men in 1980 were directly involved in entrepreneurial activity, either as employers or own account workers (Table 7.9), and though the remaining three in four were not so-occupied, entrepreneurship as a form of economic activity has become an important ethnic marker distinguishing Malays from Chinese. The supposition that Malays tend to be indolent is derived in part from their lack of business enterprise. Supposed indolence has been a source of Chinese discrimination against the Malays in employment, and this in turn has restricted economic opportunities for Malays and further confirmed ethnic boundaries and stereotypes.

Alatas (1977: 119, 151) notes that commerce was the most esteemed activity of the British colonialists, for whom industriousness was equated with acquisitiveness and direct involvement in capitalist enterprise. As a result of this view, Chinese who laboured in colonial mines and in petty commerce were considered diligent, while the Malays were branded indolent and their contribution to the colonial economy as administrators, security personnel, and as the producers of food for both themselves and urban dwellers, was underplayed. Many analysts have persisted in equating entrepreneurship with modernity, productivity, and diligence, and this has distorted their assessment of Malay society (Alatas, 1977). For example, Tham's work in *Malays and Modernization* (1983) centres on the theme that the chief 'inadequacy' of the Malay value system is the lack of stress on capital accumulation and entrepreneurship. A similar analysis is found in Betts (1975: 69), where the non-participation of Malays in entrepreneurship is the source of his assertion that Malays were marginal to the pre-1959 Singapore economy and overwhelmingly poor and rural. The assumption that Singapore Malays have tended to be rural, poor, and indolent is mistaken, as earlier chapters have shown, but the assumption has nevertheless been very important in shaping the political, social, and economic history of Malays in Singapore.

In this chapter, the reasons for the limited participation of Malays in entrepreneurship in Singapore are explored. Aspects of the internal dynamics of the Malay household which restrict the incentive and ability of Malays to engage in business were examined earlier: a limited incentive for the accumulation of great wealth due to the one lifetime view of the purpose of economic endeavour, and a practical constraint on starting a business due to the lack of pooling of capital and labour resources within the nuclear family unit. Examined here are the direct constraint on Malay business posed by their minority position in Singapore, and the social difficulties Malays encounter when doing business within their own community. The analysis explores the ways in which the cultural practices in relation to business have been shaped by the Malay cultural heritage, and by features of Singapore's ethnic and economic environment.

The Malay Niche in Entrepreneurship

The Malays and Javanese had substantial trading communities in the fifteenth and sixteenth centuries, which participated on an equal basis with Chinese, European, and other traders in the international and local trade based at Malacca and several Javanese ports. In particular, the Malays and Javanese dominated inter-regional shipping facilities (Alatas, 1977: 188). The indigenous trading class was progressively eliminated by the Dutch and the Portuguese from the mid-seventeenth century, as the colonial powers made treaties with local rulers prohibiting indigenes from undertaking international trade, or from selling local produce to the Europeans unless they were the appointed agents of the ruler. By the end of the eighteenth century, the Dutch had achieved centralized control over the entire Indonesian Archipelago and were able to establish a full monopoly over trade which they retained until the Second World War. By the time the British colonialists arrived in the late eighteenth century, they found no indigenous trading class and developed the image of the native as indolent and uninterested in profit and trade, an image which was important to the legitimization of colonial rule (Alatas, 1977: 201 *passim*).

There were a few prominent Muslim and Malay businessmen in the early history of Singapore, whose businesses were based in the area of the city known as Kampong Java. All pre-war commerce was highly segmented by ethnic group and the Malay sector focused on crafts, trades, pilgrim brokerage, the publication of Malay and Muslim texts, and food production for the Malay market. Malay metal- and leather-crafts were displaced by mass-produced imports after the Second World War. The older generation of craftsmen and traders did not diversify into new lines of business and invested mainly in land and houses. While quite a number of the present-day Malay businessmen are descendants of the older generation of traders, most started their present business afresh with their own capital and initiative.

Before the expansion of government-run training facilities in the 1970s, the access of Malays to skilled trades requiring apprenticeship was limited because of racial discrimination. Malays had something of a niche in the electrical trades, since they could obtain training from a major European company, but they were excluded from the building, plumbing, and vehicle mechanics trades which were dominated by small Chinese companies that did not generally accept Malay apprentices (Lim Kim Huay, 1960).

Cooked food is an important sector of Malay and Chinese petty enterprise. Malays have a guaranteed market, since Muslims generally do not buy Chinese food for fear of contamination by pork. Many Singapore Chinese also buy Malay food occasionally, further widening the market. Malays face difficulties in expanding their food business from the small hawker stall to more expensive restaurants because there are few Muslims who can afford to eat restaurant food on a regular basis, and because Malays like to entertain family and friends and hold ceremonial dinners at home. This preference differs from that of Singapore Chinese of both higher- and lower-income groups who give ceremonial dinners in restaurants. One successful avenue of expansion for Malays in the food business has been to develop home-catering services for weddings, where the entrepreneur provides marquee, tables and chairs, bulk food purchases, and contracted cooks. The greatest difficulty faced by Malays who are prospective entrants to the food business is the very high cost of retail space, which necessitates a large initial capital outlay and high daily turnover.

The constraint on new entrants to the food business applies generally to all new enterprises in the 1980s, Malay or Chinese. The capital requirements for business are now universally high, as urban renewal and government regulations have removed the shelter provided by backyard or unlicensed informal businesses with low overheads which formerly enabled prospective entrepreneurs to accumulate capital and experience. Collateral is required to raise bank loans, and public housing flats cannot be used as collateral. Only a small minority of Singaporeans own other types of property, fewer now than before the rehousing and urban renewal schemes. Some Chinese banks allow a guarantor to substitute for collateral, favouring those with a reputation for trustworthiness and powerful connections in the Chinese community (Lau, 1974: 22). The most successful entrants to the business world in the 1980s are the highly capitalized off-shoots of established parent companies or skill-based concerns in technical and professional lines. There are now fewer opportunities for the self-made man to work up from below than there were during the 1970s when business provided a significant avenue of mobility for many Chinese.

Businesses not requiring premises or overheads, such as sub-contracting in the ship-building, heavy engineering, or building maintenance lines, have attracted some Malays. These Malay contractors claim to be constrained by their inability to obtain contracts from Chinese firms, although they can obtain work from multinationals under European, American, or Japanese management, which are less prone to display racial prejudice. The Malay contractors find that Chinese firms give them work only when there is no Chinese bidder, and tend to give a Malay contractor unfavourable terms 'because they think we are too low; they think we Malays are satisfied if we make just a little'. Others claim to have their bids consistently undercut by Chinese competitors, convincing them that the Chinese firm is making a loss with the intention of eliminating the Malay enterprise.

Malays involved in retail have differing opinions of the possibilities of doing business with the Chinese. Some claim that Chinese wholesalers charge Malay retailers higher prices, while others find that Chinese wholesalers compete among themselves for retail outlets and offer their clients terms which are designed to develop and regularize the relationship, whether the retailer is Malay or Chinese. Some Malay retailers claim that their Chinese competitors cheat on weights and measures, so that it appears to the clientele that the Chinese offer better prices than Malays, putting the Malay out of business unless he is willing to compromise his principles and 'do business in the same way as other races'. The assumption that Chinese are often involved in cheating makes many Malays reluctant in principle to engage in business partnerships with Chinese. Language is a further constraint, and it is the lack of familiarity of Malays and Chinese in dealing with each other, rather than fear of an absolute racial boycott, which makes Malay retailers pessimistic about the prospect of gaining Chinese customers.

The perception that Chinese profit through trickery, while Malays are constrained by moral and religious scruples, is a fundamental part of the ethnic self-image of Malay businessmen. They believe that this difference in morality, which is attributed to Malays and Chinese as ethnic groups, gives their Chinese competitors an advantage over them. On the other hand, some Malays claim to obtain customers precisely because of their reputation for honest dealing, so that clients, Chinese included, actually prefer to do business with Malays, once they are convinced that the Malay businessman is competent. A long-established electrical contractor claims that a large proportion of his business is given to him by government statutory boards, whose workers are mostly Chinese:

They know they can trust me. I have never made a mistake, never grabbed an opportunity to overcharge a customer; so when people ring the utilities board and ask, 'Who can you recommend to do this job?', they always give my name. Because they know we think of our name, our mother's and father's name, and God. But after fifty years I am still not rich, unlike some Chinese after one or two years. They are brave, they take on a job beyond their means, and if they fail they go bankrupt, but if they pull through by borrowing here and there, they get very rich. We Malays guard our name first, but we never get so rich.

The practical and supposed moral constraints of doing business with the Chinese cause the majority of Malay entrepreneurs to focus on the

Malay market. At this point, the picture of Malay virtue and Chinese vice becomes more ambiguous, as the Malay entrepreneur experiences tensions in dealing with his own community. These tensions are related to ideas about interpersonal relations that are deeply embedded in Malay culture, and they are also related to the presence of the Chinese and the role of ethnicity in shaping cultural practices in Singapore.

Entrepreneurship within the Malay Community

Many conversations with Malay entrepreneurs revealed a highly charged atmosphere of distrust among traders, and between traders and the rest of the Malay community. Entrepreneurs felt that all other Malays, whether traders, customers, kin, or neighbours, were jealous of their success and were engaged in schemes to cause their downfall. Malay entrepreneurs described the attitude of other Malays towards them as jealous (*iri hati*), angry (*panas hati*), dissatisfied (*tidak puas hati*), upset (*sakit hati, susah hati*), envious (*sakit mata*), or evil (*busuk hati*). These words all refer to emotional states of mind, rather than to the economic aspects of business itself, and they show that questions of personal honour are deeply involved in business dealings in the Malay community.

Competition from another entrepreneur tends to be seen as motivated by envy and spite. One entrepreneur with an established retail outlet stated that 'if you have two Malay shops side by side, selling the same product, but one has more customers because he is more friendly, the other will be envious and upset; so he will lower his prices to sell at a loss, until both shops are bankrupt, but he will feel satisfied'. Besides competition from other entrepreneurs, Malay businessmen or women often feel threatened by gossip or rumours the Malay public may spread about them, or by witchcraft which can harm them physically and prevent them from continuing to do business.

Established traders who conduct business relatively anonymously in premises separated from their homes are less vulnerable to anxieties about their status in the Malay community than smaller traders who operate from their own home and whose market is composed primarily of friends, kin, and neighbours. However, the position of Singapore Malays as a minority ethnic group tends to make all Malays feel there is a social bond between them, so that even for the established entrepreneur there are no truly impersonal business transactions (a point to be expanded later). Another factor which increases the vulnerability of small traders is that they tend to live in low-income neighbourhoods where accusations of pride and other jealousies are more acute than in the higher-income neighbourhoods occupied by established traders who have already *achieved* social mobility.

The typical small businesses which Malays conduct from their own homes are the peddling of cooked food and the petty retail of consumer goods. This sphere is dominated by women, but some men are also involved. The following is an account of the tensions and difficulties experienced by one female petty trader living in a low-income block:

I was selling curry puffs and fried bananas from my house, then I got 'condemned', black magic, so I can't walk. The neighbours did it because I was doing well and they hate to see people better off. They don't see your hard work, only your money. The same thing happened to my sister. Her neighbours were envious of her nice possessions which were bought by her husband, who was doing well in business. So they made her ill and her husband couldn't concentrate on his business and used up his capital on doctors and medicine, so his business failed, too.

When asked about their relationship to this particular snack-seller, the neighbours denied that they were envious, and said that they did not covet other people's wealth nor interfere in their affairs. They attributed the unpopularity of the seller to her personality: she was quarrelsome and moody, and always accusing people of making her ill when doctors said she was perfectly healthy. Another cake-seller in the same block enjoyed a different reputation. She was popular before she began her business, and claimed that it was her neighbours who had begged her to start selling food because she was a good cook. Neighbours commented that 'she puts in a lot for a dollar, she doesn't take too much profit'.

The extent of the profit is one factor in a trader's popularity. Cakes and snacks have a fixed price, though they range in quality. Other goods are variable in price. Another neighbour, commenting on the unpopularity of the first snack-seller described above, believed the cause to be that the woman had overcharged for goods she had bought in Thailand for resale in Singapore. 'At first, when she sells cloth for $6, people think it is cheap so they are happy about it. But now many people have been to Thailand and they know that the price is only $1, so she says "cheap" but really she is taking too much.' The main criticism of overcharging is not that profit is illegitimate in principle but that it shows bad faith and insensitivity in interpersonal relations. The transaction is not regarded as an anonymous deal in which the trader can take as much profit as he or she is able to extract from the buyer.

Entrepreneurs adopt various strategies to minimize the tensions engendered by their business activities. One strategy is to downplay the self-serving motivation of the entrepreneur, and to claim that the business is done in order to help others, almost as a public service, as in the case of the more popular cake-seller described above. Another strategy is to underplay the extent of business activity, to claim that just a little business is done, that it is only done for fun, as a hobby, or part-time. The profit is also underplayed, and the entrepreneur claims that it only provides pocket-money for himself or herself, or for the children. Another option, particularly relevant to the provision of services, is to avoid setting a price and to claim that the recipient of the goods or services is aware of their value, and should voluntarily give a sum appropriate to the time and effort expended by the entrepreneur. This strategy shifts the onus of handling the economic element of the transaction on to the other party, and means that the entrepreneur cannot be accused of greed or lack of consideration in charging high prices. The entrepreneur may well find himself dissatisfied and resentful at how little he was given. There are expectations on both sides, and their balancing is a skilful and tense game. If one side is too calculating, or the other side insufficiently generous, the social relationship between the parties can cool. For example: 'My mother hires another person's car every morning to take food and utensils down to her food stall. Once she asked the driver to detour to collect something, and the owner charged extra. That caused a cooling of relations with the owner. My mother still uses the car, but relations are strained because she feels the owner is calculating and stingy.' In this case, the key factor in judging the appropriate price is the state of the social relationship between the parties, and the interest either party has in sustaining that relationship.

The risk of strains and tensions when business is done between parties who are neighbours, or who are in a close personal relationship, prompts some entrepreneurs to make trading at a social and physical distance their main strategy. One established trader chose to site his shop far away from areas of intensive Malay settlement because 'there are too many Malays there, they get jealous and try to put a curse on you, or say you are mean and spread stories about you'. A woman involved in petty retail was equally explicit: 'I don't sell my things here in the kampong. I have a lot of friends, especially other races, and I sell to them. Round here there are too many stories.' Other traders prefer to do business with Malays, but avoid their own neighbourhood.

Although traders may prefer to do business at a social distance, the possibility of doing this is constrained by the peculiar configuration of ethnic relations in Singapore. It was noted above that the opportunities for Malays to do business with Chinese are restricted by mutual unfamiliarity, the specialized nature of some foods and other products, and in some cases by discrimination. This situation forces Malays to look towards their own community for a market. At the same time, the strong sense of ethnic boundaries which has arisen in Singapore promotes the feeling among Malays that they are all one community, and that a social relationship exists between all Malays by virtue of shared ethnicity.

The tendency in Singapore for cultural and moral content to be imputed to ethnic categories has already been noted, and the previous chapter described how the Malay ethnic self-image has focused on the issue of the conduct of social relationships. Malays claim the qualities of consideration and humanity in relations with others to be part of the definition of 'Malayness', in contrast to the supposed Chinese characteristic of uncaring, calculated profit-seeking. This idea, which is a complex outcome of economic, ethnic, and cultural factors, has become part of Singapore Malay culture, and it enters into the decisions made by entrepreneurs. It also shapes the expectations of Malays who use the services of Malay entrepreneurs, as shown in the following statement by a Malay woman: 'I pay $100 to the Malay lady who looks after my child while I'm at work. She would charge $200-300 for a Chinese child, but Malays are considerate, and she knows I don't earn much. If I earned more, I could give her more.' The imputation of a shared moral code, generalized on an ethnic basis, constrains the Malay baby-sitter from operating as a business concern and attempting to obtain maximum profit by accepting only Chinese children who can afford to pay more. If she did this, she would be accused of greed, and she would become socially isolated from her neighbours. The assumed social, moral, and ethnic bond that links all Malays, and that is held to characterize them *as* Malays, makes it impossible for any Malays to conduct pure business transactions within their own community.

While there are advantages in trading at a social distance, there are also distinct advantages in carrying out certain kinds of entrepreneurial activities among close kin, friends, and neighbours, and this is the source of tension and ambivalence. Where a close social relationship exists between the two parties, it imposes constraints on both of them. While the seller is prevented from aggressively seeking maximum profit, the purchaser feels under obligation to have sympathy for the seller, and to purchase the goods proffered, while also feeling resentful that he or she was forced into an unwanted deal. Personalized direct selling has become very popular among Malay petty traders since retail space became so expensive in the early 1980s, and a description of the dynamics of this trade will illustrate the advantages and difficulties of doing business with people who are socially close.

The Malay entrepreneur takes goods on either cash or credit terms from a Chinese or Malay wholesaler, and resells them, again for cash or credit, to intermediate traders or directly to consumers. The goods typically circulated in this way are batik cloth, embroidered bed-linen, household utensils, house decorations, and costume jewelry. The intermediate traders are friends, neighbours or kin who 'help' the entrepreneur to sell the goods either on a commission basis or for whatever profit they can make by reselling at a higher price on credit terms. Extending the chain of intermediate traders widens the network of personal contacts that can be mobilized to make the final sale to the consumer, thus increasing the volume of trade for the wholesaler or principal entrepreneur.

In the sale of goods to the consumer, the nature of the personal relationship between the buyer and seller is often crucial. A few small traders do sell 'cold' from house to house, office to office, or in factory canteens, but a greater part of the trade is done by women on a personal basis. They visit friends, relatives, and neighbours, and go through the rituals of a social visit. They are offered a drink and they make enquiries about the family. They then produce their selection of goods in kind or from catalogues and samples, and on each of the numerous home-sale visits observed, a purchase was made. The hostess feels obliged to buy, out of courtesy to the seller who has taken the trouble to come to the house. The seller, on her part, underplays the business aspect of her visit by claiming to have brought the goods along casually or incidentally, or at the request of the hostess who has expressed an interest, or in order to help a needy friend who is trying to make a living by selling these goods. The seller emphasizes that she intends to go straight home after the visit to care for her family, that is, she is not making a round of sales calls. In private, the purchaser will state that the seller is doing business and trying to make money, and claims that she only bought the goods out of sympathy and generosity though she may also give a hint of resentment. The seller plays a precarious game: the social relationship will ensure a sale, but if taken too far, too often, or incorrectly played, the social relationship itself could be at risk.

A development of this petty personalized retail is the 'party' system which is extremely successful in Singapore but occurs exclusively among Malays. The party hostess invites friends, relatives, and neighbours to her house where she provides food, and displays the goods she has obtained from an agent. She is given 10 per cent commission on the sales, which often amounts to $400-500 for her day's work. The guests feel obliged to make a purchase, even if it is the smallest token item, and for this reason some Malay women try to avoid involvement in a social circle which often holds sales parties because they may not be able to afford the goods or wish to spend their money in this way. The guests say they would feel ashamed or embarrassed to leave without making a purchase, since the hostess has provided food for them, but they also complain in private if all the goods were thought to be too expensive. For the hostess to retain her social credit, she should ensure that there are a few cheap items for those without much money. Although Chinese neighbours or friends are sometimes invited to these parties, the Malay hostesses and a home-sales agent confirmed that Chinese do not seem to feel the same obligation to buy, and if the goods do not interest them, they leave with excuses.

Though the hostess is guaranteed a good profit, there are costs and risks. She will feel obliged to attend in turn all the parties held by her guests, and she will attempt to make purchases of equivalent value, on the principle of balanced reciprocity. She cannot hold parties too often, or her guests may become reluctant or be short of funds, and she has to allow others in her social circle the opportunity to hold parties in their turn. To preserve social credit and retain social relations which are valued in themselves and which can be used again in the future, a delicate balance must be maintained. Business profit can be pursued and everyone knows that this is the real basis of the activity, but it cannot be pursued undisguised nor to its fullest potential limits.

In the Malay view, precise and explicit calculation, or the blatant pursuit of maximum profit with no consideration for the other party, negates social relationships which are developed between individuals on the basis of a mutual willingness to 'give'. A failure to balance the demands of self and other will lead to unsuccessful entrepreneurship. It will also damage social relations and cause a loss of self-esteem on the part of the entrepreneur who, like other Malays, bases his or her claim to be a good and respected person upon the quality and extent of relationships with others. Entrepreneurial activities make the individual more vulnerable to fears and anxieties about personal relationships, and the entrepreneur feels most threatened by kin, neighbours, and friends, who are the people whose support and acclaim are most valued. These tensions can constrain entrepreneurship and perhaps discourage some individuals from starting a business, although it is widely known that there are many forms of business accessible to Malays in Singapore which can produce good profits.

The practices and attitudes towards entrepreneurship described here are the outcome of ethnic, economic, and cultural processes in Singapore which increase the difficulties for Malays to do business either among themselves or with the Chinese. There are some ways in which the cultural framework of Singapore Malay life has been used to systematically *promote* the volume of trade, as in the party sales and other forms of direct selling pursued exclusively by and among Malays, but it has also been shown that there are restrictions on the extent to which individual Malays can expand or intensify these forms of entrepreneurial activity.

The set of cultural ideas and practices relating to entrepreneurship among Singapore Malays described here is historically unique, in that it has arisen as a creative response to a particular set of ethnic and economic conditions. Cultural ideas about the legitimacy of self-interest and the importance of relationships with others have been applied by Singapore Malays in new ways and with novel effects. Some of the deeper historical roots of these ideas in the Malay cultural heritage were discussed in the previous chapter, and it has been argued that they are related to the question of individualism discussed extensively in the first part of this work. A brief review of accounts of entrepreneurship elsewhere in the Malay world will help to indicate both the cultural continuities and the unique circumstances that are reflected in contemporary Singapore Malay business practices.

Social constraints on entrepreneurship have been noted by other writers on South-East Asia, although the ethnographic literature does not contain detailed accounts of the balancing of individual rights and interpersonal relations comparable to that given here. The typical rural business is the village provision shop. This type of business is acknowledged by Singapore Malays to be the most difficult to manage, because of excessive demands for credit on basic subsistence goods such as rice and oil by co-villagers, neighbours, and kin. Malays themselves think it wiser for the Malay entrepreneur in this situation to deal only in luxury items, such as cigarettes, sweets, and soft drinks, where sympathies are not evoked. The difficulty does not stem from the illegitimacy of the business as such, nor from any suggestion of an ethic of shared poverty or village levelling such that the entrepreneur is pressured into giving away the resources that raise him above his fellows. It is simply that if he fails to give credit to people close to him when he can apparently afford to do so, he will be criticized for having no regard for his customers as *people* or for his personal relationship with them.

Nagata (1979: 112, 113) writes that in the urban kampongs of Penang, Malay traders become 'entangled in personal and social obligations' to customers and are forced to over-extend themselves in credit while their Chinese competitors are impervious to 'local social custom'. The Malays exacerbate their difficulties by doing business within the area of residence where they 'often fail to observe the cardinal rules of business in separating commercial from private relationships'. It has been argued here that, from the viewpoint of deeply embedded Malay ideas about the self and other, complete separation of commercial and private relationships is actually impossible. All social relations, including the closest relations within the household, are acknowledged to have an economic or commercial aspect, as they involve the negotiation of individual interests. At the same time, in Singapore, all commercial deals between Malays take place between people who have a form of prior social relationship, imputed on the basis of shared cultural traits and common ethnicity, whether or not they have met before. There are similarly strong ethnic boundaries between Malays and Chinese in Penang, and although Nagata has described these, she has not analysed their impact on cultural practices at the level of daily life or on specific forms of activity such as entrepreneurship. Lack of detailed studies of these issues in Malaysia makes comparison with Singapore difficult.

Other accounts of rural Malaysia describe entrepreneurs as the subject of hostility (Wilder, 1982a: 112) and bemusement for their apparent obsession with money and profit (Banks, 1983: 119). In Bailey's (1983: 80) account, entrepreneurs are described as socially integrated with other villagers, although they are more prosperous. Bailey (1983: 116, 119) found that Malay fish dealers operate within a set of social constraints in which the fishermen feel obliged to sell to a particular dealer on the basis of kinship and friendship, while the dealer in turn feels obliged to continue to purchase off-season even when it is uneconomic to do so, because of a sense of responsibility towards the fishermen. This finding is not incompatible with the analysis presented here, and it shows that with careful balancing, both personal and commercial relationships can be sustained.

Clifford Geertz's analysis of trade in Java describes two contradictory principles. He claims (1963: 40) that petty traders are under a 'normative obligation to let others in on a good living' by inviting them to join in profitable deals. This is related to his argument about shared poverty in Java, a concept which was earlier found to be misleading (see Chapter 8). Clifford Geertz (1963: 46) also claims that commercial relations are separate from social relations with kin and friends and are fully impersonal, calculating, and rationalistic; it is 'a commerce free from the constraints of diffuse cultural norms. Money is money, even between father and son, and there is no social prejudice to inhibit rational calculation.' It is rather unclear in Geertz's account whether these principles apply to different groups of traders, with only the petty traders and peasants following indigenous Javanese traditions (the *abangan*) being constrained by an ethic of shared poverty, while full-time, established traders who identify themselves more strongly with Islam (the *santri*) are free to make narrowly economic calculations. Apart from religious orientation, the particular context of economic transactions is an important factor in Java, just as it is in Singapore. Dewey (1964: 237-8), studying petty traders in the same town as that studied by Clifford Geertz, found that within the crowded urban bazaar, economic relationships were highly impersonal and transitory, while in the villages from which the traders originated, mutual social and economic obligations were important.

The interpretation of the social constraints on Malay entrepreneurship suggested by observations in Singapore is that the cultural framework of individualism recognizes few purely economic or purely social relationships. Most material transactions involve a social element best conceptualized as a gift. Material transactions are used to build and sustain social relationships, while, at the same time, social relations are the vehicle by which personal self-interest is served. This accounts both for the difficulties faced by Malay entrepreneurs, and for the peculiar combination of generosity and calculation which was described within the family and which often characterizes the relationship between the individual and others in the Malay community.

This chapter has discussed the practical and social constraints on Malay participation in entrepreneurship in Singapore, and an earlier chapter discussed constraints deriving from the pattern of relationships within the family. The direct effect of these constraints has been to close off a potential avenue of economic mobility to Malays. With low incomes, they are confined to the socially intense neighbourhoods where the constraints on entrepreneurship, and the difficulties in obtaining advance through education, are most severe. By failing to engage in business, Malays reinforce the Chinese image of them as lazy, incapable, and uninterested in economic gain. This, in turn, contributes to the discrimination by Chinese against Malays which further restricts the opportunities for Malay economic progress through either business or employment. These are some of the unintended effects of a unique set of cultural practices, which contribute to the processes of class differentiation, the reinforcement of ethnic boundaries, and the growing economic gap between Malays and Chinese that have characterized Singapore's post-1959 development.

The process of economic and ethnic structuring in Singapore does not emerge only from the practices of Malay family and community life, but also from the practices of the Chinese. If not for the Chinese predilection for entrepreneurship, the Malays would not have acquired a reputation for being non-entrepreneurial, nor would their lack of participation in business have put them at a disadvantage as an ethnic group in the competitive framework of the national economy. Singapore Malay culture has been deeply affected by the numerical presence of the Chinese, and by the real or supposed cultural differences between the Malays and Chinese as ethnic groups. The next chapter turns to a discussion of the Singapore Chinese community, focusing in particular on the questions of household relationships and the organization of entrepreneurship, since these are the two areas which appear from the viewpoint of both macro-level statistical, and micro-level cultural, analyses to be of crucial significance to the position of Malays in Singapore.

10 The Competitive Context: Chinese Householding and Entrepreneurship

THE analysis of selected trends in the Singapore economy in Chapter 7 suggested two areas in which the participation of Malays and Chinese in the national economy differ significantly. One is the labour-force participation of women, since Chinese women, especially unmarried girls, joined the formal labour force earlier and in larger numbers than their Malay counterparts. The second area of difference is the extent of participation in entrepreneurship. It was argued that these factors were in part responsible for the higher incomes enjoyed by Chinese households when compared to Malays, especially during the 1960s and 1970s, and hence contributed to the emergence of an economic gap between the Malay and Chinese communities.

This chapter discusses aspects of the Chinese household and entrepreneurship relevant to these themes. A detailed study of the contemporary Chinese family in Singapore parallel to the work undertaken here on the Malays is unfortunately lacking. The account of the Chinese family and of Chinese entrepreneurship presented here is based on a range of secondary sources, and because of the limitations of the data, the discussion in this chapter is intended to present a plausible overview rather than a definitive analysis.

The Chinese Household

Contrary to some popular ideas, the typical household in rural China was not a large extended family. The family acted as a fully corporate economic unit only while unmarried or married sons lived and worked together with their father on undivided property. On the death of the father, or often before then, pressures from the sons and their wives would cause a formal division of the property and the establishment of separate elementary families with no further mandatory economic links. Division of property was generally equal between the sons (Freedman, 1957: 19, 90; 1979: 258). The elderly would be taken care of by one of the sons, forming a stem family unit, and though the seniors retained nominal leadership, there was usually a gradual devolution of power and control to the younger generation. 'Filial piety' is the expression of formal respect for parents and their care in old age. It does not entail the economic dominance of fathers over sons unless the property remains undivided. Rich families were often more successful in delaying division of the estate, while poor households were elementary or stem in form for most of their domestic development cycle (Freedman, 1979: 305; Yee, 1959: 12).

There have been some fundamental changes in the Chinese family in Singapore. While Singapore was under British rule, and also since independence, Chinese have been subject to English family law with only minor variations related to Chinese customary law. Under English law, Chinese sons and daughters were given an equal right to intestate inheritance. Secondary wives obtained equal status with primary wives and were given a share in the estate. Sons could be disinherited because the entire estate could be disposed under will (Freedman, 1979: 140-60). These changes disrupted the legal basis of the corporate family economy. In China, sons automatically placed their labour at the disposal of the family farm or business, and they were assured of a return at the time of division. There was no distinction between working for their family or for themselves. This is now not legally the case in Singapore.

Singapore Chinese households are predominantly elementary in form. There is a remarkable similarity in the household composition patterns of Singapore Chinese and Malays. For both groups, 5-6 per cent of households comprise parents and married children, 4-7 per cent of households comprise married couples with one parent, and 72-73 per cent are elementary households (see Table 2.1 and and Census 1980: VI.45, 46). Although one study found that about 50 per cent of Chinese expect to live with their children in old age compared to 35 per cent of Malays (Chen *et al.*, 1982: 35), this difference in expectation is not reflected in the statistical trends of household composition. The government has been concerned to encourage or even enforce filial piety in Singapore, and it fears that neglect of the old is becoming a serious problem. In the 1980 census, there were reported to be 13,723 households composed of couples or widows or widowers aged over 50 years living separately from their children (Lee, 1982a: 18). If neglect of the old is a problem in the Chinese family, its source probably lies in the changed distribution of wealth and power between the generations, which is linked to the wage economy and the superior education of the young. Part I discussed the importance of these factors in the case of the Malay family, and their effect on cultural practices and household form.

Although the Chinese have a low rate of divorce with 1-3 divorces per 100 marriages in the period 1960-78 (Wahidah Jalil, 1981: 2), there are a number of families deprived of a male bread-winner due to other factors, such as death, retirement, desertion, concubinage, or paternal vices (gambling, drinking, opium), and it is possible that the Chinese family is not very much more stable in this regard than the Malay. In Riaz Hassan's study of lower-income Singapore households (81 per cent of which were Chinese), 18 per cent of households were headed by females (Riaz Hassan, 1977: 39, 40).

Relationships with kin outside the household differ greatly from those in rural China. Most early immigrants did not have close patrilineal kin in Singapore and their life was 'generally remote from kinship' (Freedman, 1979: 65). There were dialect, clan, and regional associations and secret societies which filled some of the supportive roles that would have fallen to patrilineal kin in China. Many of these institutions were devoted primarily to the provision of funeral benefits. Some associations helped early migrants with initial accommodation and a foothold in an economic enterprise or sector controlled by a particular group, but this system involved exclusion, monopoly, and exploitation as much as co-operation and mutual aid (Carstens, 1975; Mak, 1973; Freedman, 1957 and 1979).

Several studies of the contemporary Chinese population in Singapore have found that relations with kin outside the household are often as close with the wife's kin as they are with the husband's.[1] Individuals and elementary families exercise choice over which relatives they prefer to associate with more frequently. One of the most significant economic bonds outside the co-resident household is that between mothers and their working daughters for whom they perform baby-sitting services (Salaff and Wong, 1976: 34, 37; Winifred Tan, n.d.).

Economic relations between parents and married children tend to be corporate only if parents and children are involved in a common economic enterprise such as a family business and they also co-reside. In this case, the married children are given only an allowance for their elementary family needs, and share consumption with their parents. This arrangement is nearly always patrilocal (Ann Wee, 1963: 381). Instances of this arrangement have been much reduced by resettlement to high-rise flats, since the younger generation appears to have favoured living separately from parents when given the option at the time of resettlement (Chua, 1983a). Once they have established a separate residence, married children need a full salary rather than an allowance from the family business in order to maintain themselves. There is no evidence of married children working for individual wages retaining a joint economy with their parents. If they share accommodation, the children contribute part of their wages to a minimal joint budget for household expenditure, much as Malays do (Winifred Tan, n.d.).

Turning now to economic relations within the elementary unit of parents and unmarried children, it is interesting to note that women are not fully integrated into the corporate elementary family economy. In China, according to one authority, while the entire earnings of sons before the division of the estate were controlled by the father, unmarried daughters were allowed to keep their earnings from crafts and other sources. These were their personal property, and they added these sums to their dowry which became their personal property on marriage (Martin Yang, 1967: 336-40). Although as wives and daughters they laboured without pay on household tasks, or the family farm, or business, their earnings from outsiders or interest on their property were their own (Freedman, 1979: 258). Though it was legally recognized, this type of property was not encouraged by the husband's family, as it could lead the young wife to try to hasten the division of the family estate (Martin Yang, 1967: 340).

In Singapore, a study of a Chinese farming community found that wives keep control over their own earnings (Winifred Tan, n.d.) and another study of urban families confirms this (Salaff and Wong, 1976: 26, 43). It is suggested, however, that wives, especially when living with the mother-in-law, are discouraged from giving money from their personal earnings to their own mothers, indicating that they do not have complete independence to allocate their own resources as they choose (Salaff and Wong, 1976: 44).

It is not clear from the evidence available whether Singapore Chinese wives exercise their apparent right to their personal earnings by using or saving them for their own purposes, or whether they place them unreservedly at the disposal of the elementary family as a unit. The lack of a detailed study of the economic position of Chinese women as wives and daughters in the elementary family is regrettable, since this is a key issue on which comparison of Chinese and Malay practices is needed. The labour-force participation of women potentially has a great impact on the economic status of the household and its prospects for social mobility, but this depends crucially on whether or not their wages are placed at the disposal of their families. In the Malay case, it was shown that wives tend to keep their earnings to themselves, and also retain as their personal property the contributions given to them by their children, especially their daughters.

Chinese unmarried girls have had a higher labour-force participation rate than Malays since at least 1957, and the potential contribution of unmarried girls to the Chinese household economy is most significant. In 1957, over 30 per cent of Chinese girls aged 15-26 were working, and by 1980, 80 per cent of Chinese girls aged 20-24 were in the labour force (see Table 7.5). From the median age of marriage for Chinese girls, which was 22 years in 1956-60 and approximately 25 years in 1982 (Chang *et al.*, 1980: 65; *Economic and Social Statistics of Singapore*, 1960-82: 23, 24), it can be assumed that the great majority of these working girls were unmarried.

Since there is no detailed study of the relationship between Chinese working daughters and their families in Singapore, it is helpful to turn to the detailed study of this relationship by Salaff (1981) in Hong Kong for material that may be relevant. Like Singapore, Hong Kong is a Chinese immigrant society, highly urbanized, with both Chinese-owned enterprises involved in services and manufacture, and a multinational sector focusing on manufacture and assembly for the world market, employing young female labour. The similarities between the two locations do not guarantee the applicability of Salaff's Hong Kong study to the situation of the Chinese family in Singapore, but her material can provide some useful suggestions.

Salaff found that the labour of unmarried Chinese daughters in Hong Kong is a resource monopolized by their families. The Hong Kong Chinese elementary family behaves as a corporate economic unit, planning and sacrificing for its vertical mobility. The girls, who work for wages, give at least three-quarters of their pay to their families to meet household expenses and to provide education for younger siblings. The girls themselves are not the object of family investment and do not expect to benefit from the family wage economy through education or inheritance as their brothers do. One-third of Salaff's respondents lacked a father at home, but even in those families where the father was present, the wages of a sole blue-collar worker did not suffice to sustain the family. The daughters augment the family budget by 30–70 per cent, and they are encouraged to delay marriage to prolong the period during which they can contribute to the family economy. While the percentage of their wages given over to their families varies with sibling order and with the ratio of wage-earners to dependants, the daughters generally keep only a small 'allowance' for themselves, which is an amount agreed upon by their families. Out of this allowance, they save for their future marriage abode, or try to upgrade their truncated education by attending night-school. Salaff (1981: 271) concludes that, far from being emancipated by their individual wage, the daughters remain bound to a pattern of Chinese familism in which their contribution to the ongoing family unit is 'unquestioned'. She says (1981: 259): 'Ironically, wage labour force participation has enabled close-knit families to reintegrate their daughters of toil by incorporating their earnings and experiences as a promising means of attaining primary family goals.'

The Chinese parents' idea that they have a right to the child's earnings is based on the principle that children incur a debt for their birth, sustenance, and upbringing. It is this sense of debt which motivates the girls to hand over a large portion of their wages to their family. Salaff (1981: 4) cites a study of Taiwan where girls state that they give money to their family 'to repay the cost of their upbringing'. The daughters' sense of debt is particularly acute since daughters in the Chinese household are traditionally regarded as a burden, or 'goods on which one loses' (Freedman, 1957: 65; Chung et al., 1981: 15), because their upbringing incurs expenses on which there is negligible social or financial return. The opportunities for unmarried girls to work in the urban wage economy have allowed Chinese families to redeem their 'loss', and also enabled their daughters to relieve their sense of debt. Families maximize the rate of returns from their daughters by monopolizing their wages during the period between the commencement of wage-earning and marriage, and pressures from the parents, or the girls' own desire to help their families, act to prolong this period of contribution. These factors are probably responsible for the rising age of marriage.

There is fragmentary evidence suggesting the relevance of Salaff's analysis of Hong Kong for the Chinese family in Singapore. A study of female factory workers in Singapore (who were mostly Malaysian Chinese) found that the girls had a strong sense of debt towards their families for their upbringing and their very existence. This was one factor that motivated them to join the labour force to repay this debt in tangible form. Just as was shown earlier for Malay girls, Chinese girls felt it was better to give cash to their families rather than to give their unpaid domestic labour, and their ability to contribute to their families enhanced their sense of self-worth. Besides the family motive, Chinese girls, like their Malay counterparts, were also attracted to the factories by the lure of luxury consumer goods and the search for friends and companions, and possibly boyfriends or husbands (Lim Guek Poh, 1974: 23-5).

In a study of the value of children in Singapore, it was found that Chinese girls 'are valued for . . . short range utilitarian purposes' (Chung et al., 1981: 42). An interviewee in the study commented (p. 33): 'I think girls and boys are now the same. Girls go out and earn money just like boys. . . . When girls can earn money they can also give [it] to their family; [people] don't need to depend only on boys.' The study states (p. 39): 'Many of the female factory workers have to subsidize their parents' income. Some of them are forced to terminate their formal education early so that they can work and help support their family. Naturally, they do not perceive sons as the sole provider of financial help to parents.'

Although daughters are valued for their intensive cash contributions at a particular stage in the family cycle, it is to sons that Chinese families look for their long-term economic advancement and the security of the parents in old age. The study quoted above (Chung et al., 1981) has extensive material on the expectations Chinese have of their sons. Sons are the object of long-term investment through education, which is seen as advancing the family as a whole. The notion of filial piety places sons under a definite obligation to provide old-age support to their parents. There is less stress on obtaining immediate financial return from sons through their labour at a young age and it is likely that many boys make fewer financial contributions to the family before marriage than their sisters. Part of the reason for this may be that, as in the Malay case, Chinese families are not always successful in controlling the activities of their sons, and find themselves unable to prevent the youths from spending a large portion of their pay on themselves. This is noted in a quotation from a short study of the Singapore Chinese family by Ann Wee (1963: 387): 'A young man has more to spend his money on, it's my daughters who give their mother the greater part of their wages.'

According to the Chinese idea of filial piety, boys are considered to be the proper providers of old-age support to their parents, but the trend in Singapore Chinese families for sons to give more emphasis to economic and emotional relationships within their own elementary families after marriage, and the weakening of economic links between the generations caused by the wage economy, are reflected in the insecurity of Chinese parents over the role of sons. Daughters are now expected not only to contribute substantially to the family economy before marriage, but also to retain a close emotional bond to their mothers, which may include some financial aid, after their marriage (Salaff and Wong, 1976: 48; Salaff, 1981: 35; Chung et al., 1981: 42). Parents perceive the weakened relationship to sons as an outcome of the son's attachment to his wife and children, and comment, 'If I was in trouble, it is to my girls that I'd look for help; my sons are so absorbed with their wives and children' (Ann

Wee, 1963: 387), and 'when sons marry they follow their wives' (Chung et al., 1981: 42).

This section has considered the material value of Chinese children to their parents in the short and long term. No doubt Chinese parents also value their children for non-economic reasons, but the stark terms in which the expectation of direct material return from children is expressed in the Chinese family is in marked contrast to the way this relationship is perceived in Malay society. Chinese parents stated to Chung et al. (1981: 35, 38, 39):

If I have to choose between two boys or two girls, I will take two boys. This is because they are more valuable, that is, they have the capacity to earn money.

I do expect my sons to earn some money for me when they grow up [and] I sure hope my sons will help me in my old age.

I think the more boys in the family the better it is because boys generally earn more than girls and they would help the parents in the future.

The expectation of a flow of wealth from child to parent in the Chinese family is phrased in terms of the redemption of losses and the repayment of the debt of upbringing, and in terms of filial piety and the obligation to support the old. The expectation is not stated in terms of an accumulation of wealth or profit from children, although this possibility is not excluded (see Yee, 1959: 50). The children are constantly reminded of the debt they owe their parents, and this is one source of the authority parents exercise over children. According to a case study of a Singapore Chinese family, the children feel they *owe* compliance 'to their parents in return for what they receive from them, such as their food, clothes and home. It is a compliance born of dependence' (Riaz Hassan, 1977: 171). In another case study (Riaz Hassan, 1977: 173), a Chinese father's attitude towards his children is described as 'one that seeks to redeem losses incurred. For instance, when his daughters failed to complete their education, they were encouraged to be useful at home; similarly if his sons fail, then they would be better employed in some activity that would be of direct benefit to the family.'

In Singapore there are probably significant variations in the pattern of Chinese family relations according to the economic, educational, and religious background of the household. Lower-income families have more children than higher-income families (see Table 7.8). This suggests that, as in the Malay case and particularly in the older generation, children are of greater value to the lower-income group, for whom old-age security is a more significant concern. For the Chinese, children also represent a possibility for family advancement, whether through their immediate contributions to the family fund or through the provision of education to a talented child who might make even greater contributions to the parents in the future and raise the status of the family as a whole. The very low birth rate among the better educated, higher-income Chinese (and Malays) may in part reflect their assessment that children have little economic value to them, and might disrupt their career plans. The parents' class status is put at risk by children who fail academically, so emphasis is placed on a small family and very intensive concentration of parental resources and time on the children's education.

There is a marked lack of secondary data on the relations between parents and children in upper-income Chinese households. A very small sample of five unmarried Chinese professional men and women aged 25-35 were interviewed, and though the generality of the observations would need to be confirmed by a more substantial study, the findings are interesting enough to mention here. In each case, regardless of whether or not they lived with their parents, the young adults were giving about one-third of their substantial salaries to their mothers. They did so even though the parents were economically secure (in some cases millionaires) who did not need the money either for subsistence or in order to raise the social and economic status of the family unit. In one case, a son was returning regularly to his mother 30 per cent of the salary he earned in the family's own business. The interviewees stated that their parents expected money from them as a demonstration of their filial piety and formal respect. Token sums would not be acceptable. If they did not give money, their parents would criticize them for neglect and ingratitude, and employ various forms of emotional pressure to reassert their parental rights. Variations in this pattern according to factors such as the linguistic orientation of the family (Chinese or English); the parents' level of education; the family's religious beliefs (Christian, Chinese Buddhist, or Confucian); and other factors would need to be investigated.

The material presented here on Singapore Chinese household relationships suggests that while both Chinese and Malay parents have expectations from their children, they are expressed in significantly different cultural idioms. The Chinese stress a clear-cut obligation of children towards parents, which is engendered by a debt, while the Malays view the relationship in terms of more diffuse obligations engendered by gifts. In both Malay and Chinese households, notions of kinship derived from the deeper cultural heritage, have been reassessed and applied in new ways to the practical circumstances of Singapore life. In some ways, Malay and Chinese householding patterns in Singapore have become more similar in recent years. Malay children have begun to contribute to their parents' economy before marriage and in their parents' old age in a way which was not expected in the rural Malay household. This pattern of contribution to parents resembles that of the Chinese, except that it is perceived and managed in terms of a different cultural logic. For the Chinese, in the absence of a substantial inheritance in land or other forms of capital, or in cases where children have a good education and relative affluence, parents do not have any practical means of enforcing control over children and securing a share of their earnings or support in their old age. Because of this, the Chinese household in Singapore has begun to resemble the Malay system in which less emphasis is placed on clear-cut axioms and obligations, and the hopes and fears of parents are focused on their emotional ties to their children. Like

Malays, Chinese have come to rely on their daughters both before marriage and also in old age, yet Chinese still place much greater conceptual significance on the role of sons. Chinese use the 'debt' of a daughter to assert a right to her wages whereas in the rural Chinese economy, control was exercised over the labour of sons, and was linked to their right to inheritance. Although the structural economic conditions of the urban wage economy in Singapore which prompt these changes in the Malay and Singapore households are common to both groups, there remain very significant differences in the way they organize and perceive their household relationships. These differences between Malay and Chinese households are related to the distinct cultural heritage of the two groups, which has been revised and restructured, but not eradicated or changed in any uniform, predetermined way by the economic conditions prevailing in Singapore.

A significant economic impact of the difference between Malay and Chinese households identified here, concerns the contributions of unmarried children, especially girls, to the family budget. It was shown earlier that in the Malay household there is a minimal pooling of resources, and a large number of boys give no money at all to their parents, while the girls, who contribute more regularly, usually give less than half their pay. There is no quantitative data on the contributions of Chinese children to their households in Singapore, but from available data on the cultural logic at work, it seems likely that boys and almost certainly girls give much more substantial contributions to their parents, strengthening the economic position of Chinese households relative to Malay households.

One way in which Chinese households apply the resources collected from working children is in support of the education of their siblings. The corporateness of the elementary family economy, and the authority of the parents in allocating resources, makes a strategy of differential investment in children possible in the Chinese family, while in the Malay case it is problematic. Chinese girls finance the education of their brothers, and older siblings finance the education of younger ones. A study of a rural Chinese settlement in Singapore provides an example of brothers voluntarily financing the education of one of their number for the enhancement of family prestige (Chang, 1960: 41). Such an occurrence is possible, but not really expected in the Malay case, where siblings maintain separate economies both before and after marriage and do not feel linked by an idea of family name and honour.

It is possible that Chinese parents exert a stronger moral pressure on their children to succeed in school, because of the sacrifices made for them and because of the definite expectation of a future financial return on investment. There is some evidence for the role of the Chinese family system in promoting educational performance. A Chinese schoolteacher in an interview claimed that 'education ... in Chinese families is considered a part of the family fortune' (*FEER*, 31.7.81: 56), an expression never heard in Malay families, although Malay parents *do* say 'I have no material wealth, but your education is the wealth I am giving you'. An

American researcher states that Chinese and Japanese children in California obtain higher test scores than Anglo-American children, 'but only as long as their home life is fairly traditional. As they acculturate to American ways, they sink down to American IQ levels' (*FEER*, 1.3.84: 3). This suggests that the pressure to achieve *for the family* can be a stronger motivating factor for educational attainment than either American or Malay individualism.

The independent effect on education, of cultural factors unique to Malays or Chinese, should not be exaggerated, since, as argued earlier, educational performance is very strongly related to class factors and the structure of advantages and opportunities in the national economy. Nagata (1979: 127, 142, 240) found in Malaysia that Malay parents are very attentive to their children's education, invest heavily in it, often at great sacrifice, and according to research data, Malay children score higher 'motivation' levels than Chinese in the national Malay and English schools. The real opportunities for social mobility which are available to Malays under the Malaysian government's policy of positive discrimination, and the scarcity of public sector jobs and opportunities for higher education for Chinese, has clearly had an effect on educational performance.

The importance of the class position of a Malay household in determining the efforts that parents make to ensure that their children at least maintain their present status through education was noted earlier. Three studies of lower-income Chinese families in Singapore show that large numbers of children, the struggle for survival, a sense of pessimism about their children's abilities, and the influence of working peers combine in the Chinese case, as they do in the Malay, to produce failure at school (Riaz Hassan, 1977: 139; Lai, 1973; Leong Choon Cheong, 1978; *ST*, 16.5.82). The culturally validated idea that children are the producers of material wealth for the Chinese family, which could provide an incentive for parents to invest in education as a source of greater long-term benefit to themselves, is not sufficient in itself to ensure either concerted effort or success in education.

The Chinese Family and Business

Participation in entrepreneurship is the second major area of difference in Malay and Chinese economic activity which has had a marked impact on their relative position in the Singapore economy. It was noted earlier that the Chinese in Singapore have had about 20-30 per cent of their working population directly involved in entrepreneurial activities between 1957 and 1980 (see Table 7.9). Much of this business remains at petty levels, but the financial returns on entrepreneurship, even among the lowest educated self-employed, are consistently better than the pay received by employees of similar education (Table 7.10). Entrepreneurial activity has provided opportunities for social mobility for individual Chinese, and for the Chinese community as a whole in relation to Malays, both directly and through the range of employment opportunities created by Chinese entrepreneurs, which are largely inaccessible to Malays. This section examines the role of the Chinese family and community structures in facilitating Chinese entrepreneurship.

There is good evidence that small Chinese businesses do successfully monopolize the labour of household members. The wife and unmarried children of a Chinese entrepreneur labour in his business without pay as part of their unquestioned duty. This elementary family corporateness was shown to be absent in the Malay case, although there are instances of voluntary co-operation.

A study of Singapore hawkers, 98 per cent of whom are Chinese, found that the 'hawker family constitutes an economic unit...each member of the family can be pulled in to contribute his or her share in ensuring the success of the hawking trade without expectation of any material reward' (Charlotte Wong, 1974: 86; also Chang, 1960: 37). Charlotte Wong (1974: 96) found that 49 per cent of hawker households have *all* the household members involved in the business. Though some hawkers operate alone, those who employ assistants usually employ those related to them. Only 12 per cent of hawkers had assistants unrelated to them who were paid a fixed wage, while 62 per cent of related helpers received no pay, and a further 27 per cent received either a 'token sum' or a share of the profits (Charlotte Wong, 1974: 200). Since Charlotte Wong does not specify the degree of relationship between owners and assistants, those who receive some form of payment could be individuals who are related but outside the co-resident household. The majority of hawkers are uneducated or educated in the Chinese language, and their deeper link to Chinese cultural traditions and ideas of filial piety is reflected in their household composition. Charlotte Wong (1974: 84) found that 29 per cent of hawker households have more than one family nucleus (usually parents and married children), compared with 12 per cent of the total population in 1970.

Malays state that they believe close family co-operation to be essential to the hawker trade, because of the long hours and rapid cashflow involved. Malay hawkers often complain of having 'no one to help' because their spouse, children, and other kin are otherwise occupied and they feel they cannot trust or afford to pay helpers. This often results in the Malay hawker reducing his or her hours of operation. It is noticeable all over Singapore that Malay food stalls are often closed or sold out, the owner complaining of tiredness, although there is a good market for their product.

Beyond the elementary family, or the extended family before the division of property between sons, the Chinese family system does not incorporate any other relatives into its business ventures as of duty or right. Brothers often cease to co-operate after the death of the father and division of the estate (Freedman, 1957: 90). It is common for a firm established by a father to break up after his death, or to change from a partnership to a sole proprietorship as one of the brothers buys out the shares of the others (Yang Hon Loon, 1973). The Chinese family structure, with equal inheritance between sons and their subsequent economic independence, is *not* ideally suited to the continuity or development of business across the generations. However, the parents' monopoly of the labour of unmarried children, and often also the participation of married sons in the parents' business until they receive their inheritance, do permit a prolonged period of corporate endeavour and accumulation which is totally absent in the Malay case.

Although kin outside the elementary or extended household are not automatically involved in a business venture, and have no rights or claims, many studies have shown that the Singapore Chinese prefer to operate business on the basis of reduced social distance. This results in the involvement of kin, quasi-kin of the same clan surname, co-dialect speakers, and other Chinese as employees, partners, and trading contacts in preference to unknown individuals. This practice can be traced to the history of the immigrant Chinese community in South-East Asia. In the early days of Singapore, all business contracts were verbal and were enforced informally by the Chinese community. There was a preference to do business with people who had a community reputation for reliability and honesty, and office-holders in Chinese associations had a special advantage in making contacts and demonstrating trustworthiness (Freedman, 1957: 94, 96). This way of doing business involved a strategic use of personalized relations, and was highly flexible, since a whole range of types of connection could be brought into play as immediate business needs dictated. Friends could be as important, as trusted, and as close as actual kin. It is the personalization of economic relations, rather than any obligation to incorporate actual kin in business dealings, that is focal to Chinese entrepreneurship in Singapore. This point was made by Freedman (1957: 88), who wrote that business rests on non-kinship solidarities and 'the real distinction is not between family and non-family, but between personal and impersonal'.

The pattern of kinship and business relations described here is certainly not identical to that of traditional, rural China, where patrilineal kin had real economic importance, but neither is the contemporary pattern totally unrelated to the Chinese cultural heritage. The present pattern is an outcome of the creative application of cultural ideas about kinship and trust to the peculiar situation of the Chinese as migrants in South-East Asian cities. It could be that the absence of patrilineal kin gave the overseas Chinese a positive advantage over their rural Chinese cousins, since it gave them more flexibility in business dealings. Chinese entrepreneurial success in South-East Asia can be attributed, in part, to this peculiar and unique combination of elementary family corporateness, the economic contribution of married children if the division of property could be postponed, and *de facto* flexibility to emphasize connections with sets of kin other than those defined by strict patrilineal terms, together with a traditionalized ethic of kin trust and co-operation that could be extended strategically to non-kin. By contrast, the Malay system, though bilateral, lacks elementary family corporateness and opposes the ethic of kinship (non-calculation, the gift) to that of commerce, making business within the close social circle particularly tense and difficult.²

Malays certainly attribute Chinese business success to their co-operation with one another as kin, on the basis of networks of contacts, and as an ethnic group. A Malay writer has described the alliances and monopolies that can successfully exclude non-Chinese from favoured areas of business (Mahathir bin Mohamed, 1970: 54). It is important to recall, however, that in addition to co-operation, trust, and mutual aid, business is concerned with profit. The Chinese pattern of personalized business relies on an acceptance of the view that business and profit can legitimately and indeed most advantageously occur between people in a close personal relationship, whether this relationship pre-existed or was developed for the purpose. The previous chapter made a detailed analysis of the perception and management of Malay business relationships, but regrettably there is no parallel study which would indicate whether Chinese, too, experience tension and ambivalence when conducting business among kin and friends.

The discussion here has focused on the role of the Chinese family and social organization in promoting business, and only brief mention will be made of some of the economic conditions that favoured the early entry of the Overseas Chinese community into business. One factor was their exclusion from land-ownership in colonial Malaysia to which, as former peasants, they might otherwise have been attracted. A second factor was their intention in migration which was not, originally, permanent settlement, but the accumulation of wealth overseas in anticipation of a return to China. This attracted them to entrepreneurship which was the only pursuit offering a *hope* of windfalls and the rapid accumulation of liquid assets (Linda Lim, 1983a). In the present economic environment in Singapore, the most profitable entrepreneurial role is available to those who have the capital and management or technical expertise to go into joint ventures with multinationals. The Chinese companies which have succeeded in this area have had to abandon most of the personalized 'Chinese' ways of doing business described above, but accumulated capital and competence from earlier ventures certainly gave them a headstart in the new phase of Singapore's commercial development (Linda Lim, 1983b).

Another feature favouring Chinese entrepreneurship that is related to the circumstances of migration is the emphasis on wealth as the primary indicator of personal worth. In the absence of the established Chinese administrative élite, the Mandarinate, wealth became the key indicator of success among Singapore Chinese. Since all successful Chinese men in Singapore had made their own fortunes in the space of one or two generations, ancestry became less important and individuals tended to be judged solely on their own performance. Office-holding and philanthropy were sources of prestige, but they were only possible on the basis of wealth. They did not provide alternative avenues to high status (Freedman, 1979: 64; Nagata, 1979: 159).

In rural China, according to Skinner (cited in Madge, 1974: 185), obligations to the ancestors and wider kin group provided strong motivation for economic endeavour:

The Chinese peasant had a definite place in a temporal continuum of kin. Within the extended kin groups—dead, living, and yet to be born—he looked to the past as well as to the future: he was not only grateful to his ancestors for what his immediate family had, but was responsible to them for what he did to further the fortune of his family and lineage. His world view was, thus, historical and kin centered, and in this context his industriousness and thrift served ends transcending his individual life. His primary goal was not individual salvation, but lineage survival and advancement.

In Singapore, lineage and ancestors are less important, but there is an interest in establishing family property and prominence which can be passed on to descendants.

Ultimate personal and religious values also play a role in the incentive to accumulate limitless wealth through business. For many Chinese, peace and plenty in the afterlife are directly and unambiguously related to the wealth accrued in this world by the deceased himself or by descendants who dutifully carry out ancestor-worship rituals (Freedman, 1957: 209; Alatas, 1972: 31). It was shown earlier that the relationship between Malays and their descendants over the question of wealth and obligations after death is more ambivalent. The difference in religious motivation between Malays and Chinese does not wholly determine the pattern of economic activity, however, since it was indicated that some Malays do believe that they will benefit in the afterlife from their wealth and donations to religious causes. Moreover, the importance the Chinese attach to material wealth does not alter the reality that 75 per cent of Chinese men are employees rather than entrepreneurs (Table 7.9) and the majority of Chinese are neither successful nor wealthy (Table 7.4). As Swift (1963: 241, cited in Madge, 1974: 183) points out, it is not cultural ideas taken in reified isolation, such as the notion that Chinese are interested in wealth and Malays are not, that explain Chinese strength and Malay weakness in entrepreneurship. Here, it has been argued that there is a whole range of cultural, ethnic, and economic factors involved, and the difference in performance between Malays and Chinese in this area is best examined in terms of the cultural knowledge and the practical circumstances that affect the organization of economic relations both within the household and in the wider community.

1. See Ann Wee (1963); Freedman (1957); Chua (1983a); Leong Pui Ling (1982); and Winifred Tan (n.d.) for discussions of changing trends in the day-to-day practices and organization of Chinese households. See Ward (1963) for a general discussion of the impact of the urban wage economy on the Chinese family. See Wilder (1982b) for a discussion of some differences between Chinese and Malay families.

2. Swift (1963: 243) and Dewey (1964: 237, 240, cited in Madge, 1974: 183) argue for the importance of Chinese family structures in entrepreneurship, and make the contrast with Malay and Javanese individualism. The present account has provided more detailed substantiation of their insights.

Part III
Culture, Economy, and Ideology

Introduction to Part III

PART II was devoted to analysing the impact of Chinese and Malay cultural practices on the economic and ethnic structure of Singapore society. The attempt was made to show how, through their culturally informed activities in the course of everyday life within the household and in the community, people unintentionally create those very conditions–economic differentiation, ethnic barriers, and so on–which then appear to them as objective structures of opportunity and constraint. As these conditions become embedded in the institutions of society, such as census categories, good and bad schools, high- and low-income neighbourhoods, they form the framework within which daily life is pursued. These conditions, in turn, play a part in shaping the new cultural knowledge upon which the practices of daily life are based.

This book has explored the interactions between forms of cultural practice and broader social and economic institutions, and it has been argued that neither plays a part independent of the other, nor can cultural or economic processes be understood in reified abstraction. The final part of this work looks at the way Malay culture and economic performance in Singapore have been interpreted by observers both within and outside the Malay community, and identifies a tendency to reification which has distorted the understanding of the position of Malays in Singapore and some of the significant national trends in which they are involved.

11 Culture, Economy, and Ideology

> Ha! What does the 'professor' think? We're talking about Malay progress, and why the Malays don't get on in life, until we've nearly run out of ideas. But we still haven't found the secret. Maybe you'd be kind enough to explain your own thoughts about all this. (Harun Aminurrashid, 1932 quoted in Roff, 1967: 156)

NOTED at several points in the discussion so far, was the prevalence of an image of Malays in Singapore as rural, poor, and backward. The issue arose in the context of the description of the occupation and settlement patterns of Malays in Singapore, where there has been a tendency to exaggerate the numbers involved in agriculture and fishing, and a mistaken view that urban kampongs were rural settlements. The Malay political leaders of the 1920s, who were searching for markers of ethnic identity, reinforced the rural image by asserting that only in a kampong with rural architectural style and fruit trees could a Malay really be himself (Hanna, 1966: iv; Roff, 1967: 193). The image of backwardness arose again in the discussion of the pre-1959 economic position of the Malays. The weak development of the Malay economic élite promoted a popular view that Malays were generally poor, although it was shown in Chapter 7 that the economic position of the majority of Malays relative to the majority of Chinese deteriorated only after 1959. The popular image was based on a distorted comparison of rich Chinese with poor Malays, rather than equivalent socio-economic groups. The image was also noted in the context of the discussion of business, where it was argued that entrepreneurship is viewed by many analysts and observers as the yardstick of diligence, progress, and productivity, leaving the impression that while the South-East Asian Chinese is the quintessential 'economic man', the Malay is not. This view was held by the British colonialists, and it is the view of many Chinese, forming part of their assessment that Malays are indolent.

The most common explanation for Malay 'backwardness' is their supposed adherence to cultural values that observers hold to be detrimental to economic progress. It is not the intention here to deny the effects of culture. On the contrary, this work has analysed at length both the effect that cultural ideas have on shaping day-to-day activities, and the unintended consequences of culturally informed practices on Malay participation in the national economy, and on the economic and ethnic structure of the nation itself. The argument made here has been that culture is historical. Cultural knowledge is produced, reproduced, and altered in the course of daily life as it is applied in creative ways in specific conditions. By contrast, the cultural values that have commonly been held responsible for Malay economic backwardness in Singapore tend to be viewed in a reified way as permanent and unchanging characteristics of the Malay race. Although economic factors are also recognized in analyses of the Malay position, the role of cultural values is usually given prominence and these values are seen as independent or somehow unrelated to the circumstances of daily life.[1]

An alternative analysis of the historical, material, and cultural circumstances of economic and ethnic differentiation in Singapore has been presented here, but the image of Malays as perennially backward, or reified cultural explanations for Malay backwardness, cannot be dismissed as merely false. The image of backwardness and its supposed cultural causes have themselves become part of the cultural fabric of Malay and Singapore society, and they have real practical effects as they are incorporated into the daily lives of ordinary Singaporeans and into national political processes. It will be argued here that cultural explanations for Malay backwardness have come to play an ideological role in contemporary Singapore, and this chapter considers the nature and effects of this ideology.

Adopting the position taken by Giddens (1979: 187) linking ideology to questions of the distribution of power and resources, the term 'ideology' is not used here to refer to symbolic systems in general–the term 'culture' is used for that purpose. Neither is 'ideology' used only to refer to explicit political doctrines, nor is it identified by characteristics such as truth versus falsehood or science versus non-science. 'Ideology' is defined as those aspects of symbolic systems which legitimize the sectional interests of dominant groups (Giddens, 1979: 188). The identification of a particular symbolic system as ideological, depends on the analysis of the structures of domination in society by virtue of which some interests can be identified as sectional. It also depends on the analysis of the practices and meanings of everyday life, since it is at this level that symbolic systems, including their ideological aspects, are developed, sustained, renegotiated, or distorted and where the crucial effect of ideology, legitimization, is obtained.

Reification and the denial of history is one of the principal modes of operation of ideology. By projecting present social and economic conditions back into the distant past, and tracing them to pre-given, supposedly unchanging cultural conditions, the present division of rewards in society is made to appear inevitable, naturally occurring, and therefore just. It is not argued here that cultural explanations of Malay backwardness were formulated conspiratorially to dupe people, nor imposed systematically by the government or any other party in a planned campaign of misinformation. Symbolic systems which perform ideological functions can arise from the interplay of cultural and material processes at any level of society, and they are not accepted uniformly by all sections of the population. In this case, cultural explanations for Malay backwardness have a long history. They have been generated by colonial officials, by the Chinese, and by immigrant Malays assessing the indigenous Malay population. An important source of these views has been real differences in the cultural practices of Malays and Chinese, particularly with regard to entrepreneurship. Cultural explanations of the Malay position which for some groups of the population are tacit, practical knowledge, or 'common sense', have also been taken up and made explicit and systematic by articulate sectors of the Malay and academic community. To describe this systematic form of an ideological system, Bourdieu's (1977: 169) term 'orthodoxy' can be adopted.

The first part of this chapter traces the development of the orthodoxy that Malays are economically backward because of their cultural values. Subsequent sections consider the extent to which the cultural explanation for Malay backwardness is incorporated as an element of cultural knowledge shaping public and private discourse and the practices of daily life, and the extent to which it plays the role of legitimizing domination.

Culture and Malay Backwardness: The Development of the Orthodox View

As early as 1894, an article appeared in the Singapore press on Malay weaknesses, which were then identified as adherence to custom, lack of industry and ambition, hostility to those with talent, and inability to practise self-help (Roff, 1967: 54). In the periodical *Al-Imam*, founded by Singapore Muslim reformists in 1906, constant attention was given to the cultural inadequacies of Malay society, in attacks which Roff (1967: 56) describes as 'an orgy of self-vilification and self condemnation', focusing on backwardness, laziness, complacency, and the tendency of Malays to bicker among themselves. Many of the Muslim reformists were of mixed Arab or Indian Muslim and Malay descent, but there were also indigenous Malays and Javanese in the group, and these Malays did not dispute the critical assessment of Malay society. The tendency of the Malay and Muslim élite to assess Malay society in negative terms was present throughout the 1920s, when the 'underlying conviction held by so many Malays [was] that as a people they were politically incapable, economically inept and culturally inferior to others' (Roff, 1967: 150). Similar opinions continued to be heard until the 1940s.

Alatas (1977) has analysed at length the development of the image of Malays as backward during the colonial era, and its role in legitimizing colonialism. Malays and other natives of South-East Asia became branded as lazy because their rural subsistence base enabled them to reject virtual slave labour on colonial mines and plantations. Non-Malays who were imported to meet colonial labour needs were told that their presence was required because the Malays were not able or willing to do the work and their respective roles in the colonial division of labour shaped the images the subject peoples held of each other. Alatas has pointed out the continuing influence of the 'myth of the lazy native' on non-Malays and the Malay intelligentsia, and examined in detail two Malay works which show that the theme continued to be popular right up to the 1970s. The works are Mahathir bin Mohamed's *Malay Dilemma* (1970) and UMNO's *Revolusi Mental* (Senu Abdul Rahman, 1971). This chapter will not duplicate Alatas' analysis, but rather focus on the 'Malay problem' as it has been seen in Singapore since 1959. Cultural explanations which were developed in the course of justifying the ethnic division of labour under colonialism were among the concepts available to the Singapore government, the intelligentsia, and population at large when it came to providing explanations for the failures of free social mobility in a 'meritocratic' society after 1959.

In 1965, an editorial in a publication of the Malay Youth Literary Association (PPPPM), one of the earliest associations devoted specifically to Malay educational improvement, noted that Malay children in the kampong could be seen out late at nights, playing sports and wasting time, and that neither they nor their parents appeared to value education (Saim Noordin, 1965). This theme was developed in a campaign named Gerakan Obor (light the torch) launched by the association in 1966, which comprised a series of radio lectures in which Malays were urged to change their attitudes, especially with regard to education. In these lectures, there was some analysis of the factors that impeded the education of Malay children, such as low incomes, large families, and absent or ineffectual parental guidance and control. There was no mention of the large families, low incomes, and probable late nights of the 60–70 per cent of Chinese children who also left school with primary education or less during the 1960s and 1970s (see Table 7.13). Although it has been argued here that the comparison of Malay and Chinese educational performance for the mass of the population does not appear to give cause for overall cultural indictment, Malay leaders were convinced that there were Malay cultural deficiencies in this area. The opening lecture of the Gerakan Obor campaign claimed the need to change the attitudes and ways of thinking of Malay people, to make them adapt to change, not to be complacent, day-dream, live in the past, or waste time in cinemas or nightmarkets or watching Malay dramas whose content was irrelevant to progress. Day-dreaming, complacent, and living in the past; the image is that of the lazy native of the colonial era. The 'problem' is seen as uniquely Malay, cultural, and deriving from a supposed failure to change.[2]

Several analyses of the 'Malay problem' were published in the early 1970s. There was a seminar on Malay education, which noted the real problems of discrimination, low economic value of Malay-language education, poor facilities in Malay schools especially at upper secondary level, and the economic problems faced by Malay families. It also asserted that Malay parents paid less attention to education than other races (Mokhtar Abdullah, 1968–70: 17), and that Malay children were brought up not to ask questions, were too shy to talk in class, and were indisciplined through permissive child-rearing practices. Based on this negative assessment of Malay culture, the report claimed that 'unless the ethic of Malay life is so changed to one where work–a job well done–is of great value', no education system could improve Malay performance (Gordon, 1968-70: 45, 46). Here, again, the implication is that indiscipline and a failure to value work are peculiarly Malay characteristics.

In 1971 there was a seminar on Malay participation in the national development of Singapore. There, too, analysis was made of structural impediments to Malay progress–incomes, discrimination, and Malay education, among others. It was accurately perceived that few Malays were enjoying the benefits of gracious living in Singapore's era of prosperity. It was not noted, however, that just as many Chinese were excluded from prosperity, since Malays and Chinese had a similar proportion of their population in the low-income group until the early 1970s, and a greater proportion of all Singapore's population was in poverty in 1973 than twenty years earlier (see Chapter 7). The opening address of the seminar described how Malay culture had been shaped by British colonial education policies which made Malays contented, obedient, and without inquiring minds. It suggested that, in the enclosed world of their kampong, Malays had become involved in religion and gracious living, and unconcerned about material progress. Malays were changed by colonialism but Malays, unlike the other races, had not 'modernized': 'Although the other races too had to modernize, the adjustment was a minor one in that the other races had in fact started modernizing since 1819 whereas for the Malays this did not start until 1959' (Sharom Ahmat and Wong, 1971: 3, 4). Here it is asserted that the past was basically unchanging for the Malays, though not for the Chinese, before the People's Action Party came to power in 1959.

Another analysis which asserts that Malay society failed to change before 1959, appeared in a Ph.D. thesis by Betts, an American political scientist. He wrote (1975: 19) that 'much about the Malay way of life was acknowledged to be generally incompatible with swiftly changing Singapore, and factors intrinsic to the community inhibited rapid or easy acceptance or internalization of change'. He considered (1975: 37) that the suburban kampongs were 'essentially rural' and that 'many Malays contented themselves with fishing, subsistence agriculture, and other non-urban based economic activities'. It was shown in Chapter 6 (p. 102) that Malays and Chinese had similar proportions of their population engaged in rural pursuits in 1957, 7-10 per cent, and Malays were involved in the whole range of urban occupations. Betts (1975: 43) asserts, on the basis of the same 1957 census data, that Malays were 'only marginally participant in the economic and political life of Singapore'. He concludes (1975: 50) that up to the war, there 'was neither incentive nor desire for the Malays to abandon their Islam-centric, traditional, and rural-oriented way of life for the far less pleasant life of the modern urban labourer'. According to the statistical evidence of the census, quite apart from the ahistoricism of his analysis, Bett's assertions are seriously mistaken.

Bedlington, also writing a political science Ph.D. dissertation on the

Malay community, makes a similar claim that Malays were rural and unchanging prior to 1959. He wrote (1974: 514) that Malays have '... remained isolated in residential patterns of cultural and geographic insularity, bounded by the rural *kampong* and the urban ghetto. Inside these cultural redoubts, uprooted Malay peasants, brought to the city but never really urbanized in a psychological sense, have created replicas of their rural past with which to nurture traditional values.'

According to Bedlington (1974: 515), Malays emerge daily to work with non-Malays but 'at night they return to their communal enclaves rarely admitting the relevance of these outside influences on their own lives'. While he recognized that there were also some Chinese and Indians who were poor, he argued (1974: 515) that it was only the Malay community that remained impeded by '...ethnically defined cultural inhibitions. Non-Malays have been able to make a complete break from the embrace of those traditions that retard economic development, but the Malay community continues to be affected by contradictory cultural impulses.'

For Bedlington, too, other races in Singapore have changed while Malays have remained static, without history, impeded by their cultural values. The 'values' Bedlington discusses are only those he deems negative for Malay progress, and they are described as permanent, reified, cultural traits, in abstraction from their social and material context. He argues that Malays are improvident in the spending of their income, demonstrated by their possession of radios, televisions, and other luxury goods often bought on hire-purchase (Bedlington, 1971). Although he may not have had access to this data, a glance at the 1972/3 *Report on the Household Expenditure Survey* (1974: 55) would have shown that, in each income group, Malays and Chinese spend almost identical portions of their budget on food, housing, clothing, and miscellaneous items. Bedlington (1971) also argues that the Malay idea of *rezeki*, or belief in the predetermination of man's economic destiny, results in fatalism and a 'lack of will to go on striving'. Interviews and observations in the course of this research did not find the concept of *rezeki* to be at all prominent in Malay thinking about economic matters, as it was never mentioned by businessmen, potential entrepreneurs, or workers in explaining their economic lot. There is more ethnographic evidence for the importance of fatalism with regard to personal economic destiny in Singapore *Chinese* culture (Vivienne Wee, 1976: 171, 172; Lai, 1973: 35; Leong Choon Cheong, 1978: 119), suggesting that, if fatalism is a factor affecting economic behaviour, it is not unique to Malays.

Bedlington (1971) also discusses the heavy spending by Malays on ritual feasts (*kenduri*) and weddings, not noting the economy of a self-catered Malay wedding compared to the expense of a formal Chinese restaurant wedding banquet, which is an event into which many low-income Chinese families pour years of savings. In any case, both Malays and Chinese finance these weddings by a system of balanced reciprocity through which the guests, by their monetary gifts, pay for the dinner and the roles of host and guest are expected to be reversed on future occasions. Bedlington also notes Malay 'other worldliness', deficient early socialization and other factors he considers to impede education. His (1971) comment on the structural impact of low incomes is that it 'is a significant variable–but Malay writers... have stressed that Malay students themselves must change, especially with regard to the lack of "achievement" motivation...'.

Bedlington states that 'informed Malays' were his sources on Malay values and their detrimental effect on Malay progress. The Malay reforming élite group with whom Bedlington associated, later went on to use his research as academic confirmation of the cultural causes of Malay backwardness. In this rather circular way, the Malay cultural weakness orthodoxy became more firmly established. It is not uncommon that academic literature, in this case sociological work of dubious quality, enters into the cultural scheme and hence the history of the subjects analysed. This occurs because, unlike inanimate objects of investigation, people seek to understand their own history and conditions of existence.

It must be stressed again that cultural explanations for Malay backwardness were not the *only* explanations that were given. Attention was also paid to material circumstances, such as the status of Malay education, discrimination, and observable practices in which Malays differed from the Chinese in the early 1970s such as the birth rate and female labour-force participation rate. Examples of analyses that examined these factors are the series produced by the *Singapore Herald* in 1970, and a paper by Ahmad Mattar (1979). But the stronger tendency has been to look for cultural explanations, and to state that cultural factors are the prior, real, or more fundamental, problem.

The Malay culture-weakness theme was renewed in the late 1970s under the impact of the Islamic revival among the educated Malay élite, who continued to be concerned about Malay social and economic problems. This group comprised members of the Muslim society at the university and the leaders of various other Malay and Muslim associations. They considered that Malay values, in particular incorrect interpretations of Islam, were the cause of Malay backwardness. A passive attitude towards religion which Malays saw as guaranteeing their future without the need to make an effort, fatalism, an overconcern with rituals, and satisfaction with short-term results were described as the negative, non-progressive, mistaken, and non-Islamic views that had crept into Malay religion (National University of Singapore Muslim Society and National University of Singapore Malay Society, 1982: 9). A Malay undergraduate thesis states that pre-Islamic values relying on 'magico-animistic explanations for worldly phenomenon... retard the growth of scientific orientation, self confidence and entrepreneurship [and] encourage passive resignation, self reproach and a lack of achievement orientation...' (Yang Razalie Kassim, 1979: 133).[3] A Malay government minister, when describing the positive value of Islam for Malay development, made a similar, though implicit critique of detrimental Malay values which impede progress, when he claimed that Islam stresses work and effort, forbids laziness, enjoins long-term planning and thinking about how to solve problems, and forbids waste, arrogance, complacency, and despair (Rahim Ishak, 1979).

The analysis presented here of the speeches and writings of some academics and the Malay élite concerned with Islamic reform and Malay progress, has traced the development of the view that Malays in Singapore have been rural, unchanging, and impeded from development and progress by aspects of their culture.[4] Each set of writings referred to previous ones in the same vein, and by repetition and amplification this interpretation of the Malay predicament, past and present, became 'authoritatively' established as an orthodoxy among the Malay élite. Taking the lead from some of these written sources and also from the unwritten, often inarticulate explanations for Malay backwardness deriving from the contexts of daily life in Singapore, this idea about Singapore Malay culture has become further popularized in the press, and has also entered official publications. For example, the image of the unchanging, rural Malay is present in this feature article which appeared in the national press in 1980: 'In two taxis and a jeep, we descended on the sleepy community of Andrews Avenue. The residents cast sidelong glances as they lazed like lizards in the sun.... At the sight of the camera, tripod and other paraphernalia, they jolted upright in open-mouthed wonder' (*ST*, 2.12.80).

A review of the history of Malays in Singapore that appeared in the government publication, *The Mirror*, placed a mistaken emphasis on their rural past (1.2.83) and a special publication by the government board for Islamic affairs suggested that the 'most significant' factor indicating progress and change among Malays was the drop in the percentage of Malays involved in agriculture and fishing from 1.24 per cent of the national population in 1957 to 0.19 per cent in 1980. The 'significance' of this particular trend is belied by the very statistics used to support it. The publication commented, 'To put it crudely, the boys were "getting off the farm and into the factory"' (Muslim Religious Council of Singapore, 1984).

Economic Differentiation and Cultural Orthodoxy in the 1980s

When the 1980 census was published, Malay leaders and the government were deeply concerned by the deterioration of the Malay position relative to the Chinese in both educational attainment and occupational status during the preceding decade. A government minister, Goh Chok Tong, lamented the small number of Malay graduates and professionals (*ST*, 22.8.80). Prime Minister Lee Kuan Yew pointed out in his National Day Rally speech that a higher percentage of Malays than other races had not completed their primary or secondary education, and suggested that this be given attention by Malay leaders (*ST*, 22.8.81). Another minister, Tony Tan, pointed out that Malays had a higher drop-out rate than the overall national population: 9 per cent of Malays versus 6 per cent of all races in the age group 10–14 years were not in school, and

69 per cent of Malays compared to 60 per cent of all races in the 15-19 age group had dropped out of education. He urged parents to ensure that their children went to school to benefit from the expanding education opportunities in the 1980s (*Warita*, 1982, Vol. 18: 10-11).

It was argued earlier that differences in Malay and Chinese performance in education in the 1970s were not primarily caused by cultural features unique to Chinese and Malays as ethnic groups, but by the process of economic differentiation in Singapore and inequalities in opportunity which affected all races. An official review of the national education system in 1978-9 was provoked by a recognition of the problem of early school-leaving, and it was noted earlier that educational performance was fully recognized by the government to be related to the economic and educational background of the student's home. Chapter 7 showed that by 1980, Malays had become concentrated in the lower-income sector of the national population. It is thus unlikely that poor Malay performance in education results primarily from problems internal to the Malay community, although this *appears* to be the case when ethnically based statistics compare one group to another without allowing for socio-economic status. The established Malay cultural-weakness orthodoxy predisposed government ministers to interpret Malay educational performance as a unique Malay problem, in abstraction from the national debates on education and class.

There are exceptions to the tendency to perceive Malay educational performance in cultural or ethnic terms. Devan Nair, while President of Singapore, noted several times in a speech to Malay parents that the bulk of Singapore's population had low incomes and poor prospects, Malays among them (*ST*, 20.1.82). Goh Chok Tong noted the effect of class factors and parental support on the educational performance of children without making any comments indicating that this problem was unique to Malays or Malay culture (*ST*, 19.7.84). The stronger tendency, however, was for the government to see Malay educational performance as a Malay problem, and this interpretation of the problem substantially determined the solutions proposed. Whether poor Malay economic and educational performance was attributed to specifically Malay causes, such as Malay culture, or to the circumstances of Malay history, or whether Malays were seen as merely over-represented in the lower-income group for unspecified reasons, the firm suggestion from the government was that *Malays* should seek to solve these problems. There was no suggestion that Malay educational performance was directly related to the overall structure of the national economic and educational system which made mobility through education a scarce prospect for the great majority of Singaporeans. In the government view, the primary cause of the problem lay within the Malay community, and this indicated to the government that cultural changes within the community would be the most appropriate solution. There had to be changes in Malay beliefs, life-style, and attitude to education and work. The Prime Minister (1982b) stated that no government efforts could help the Malays as effectively as could the Malay leaders themselves, since only Malays would be able to propound and instil the necessary set of new and progressive values in the community.

The Malay leaders to whom the government looked to solve these Malay problems were the Malay Members of Parliament, the officeholders of Malay and Muslim associations, journalists, teachers, religious teachers, lawyers, and businessmen who comprised the section of the Malay economic and educational élite concerned with Malay reform. The 1980 census had awakened the same concerns in these Malay leaders as it had in the government, and they took prompt action by forming an organization devoted to improving the level of education in the Malay community. The formation of this organization, named Mendaki, marked a new sense of urgency in the perennial efforts at Malay reform. Its understanding of the nature of the 'Malay problem' was, however, firmly based on the orthodoxy of Malay cultural weakness. Mendaki's analysis of the Malay problem was set out at the 1982 Mendaki Congress. The collection of Congress papers begins with this quotation from a well-known Malay intellectual, Zainal Abidin bin Ahmad, who wrote about Malay problems in the 1940s:

We Malays are generally a poverty stricken people. That is the clearest and most thought-provoking character of our race, a deficiency which makes us lose out, or at least stay backward in the march for progress. Poor in terms of education and training, poor in terms of money, poor in desire and ambition, poor in brain power and poor in that quality of high and honorable character–no wonder we are mired down and backward in the road forward (when compared with other races)....[5]

In choosing to use this quotation, Mendaki confirmed that it viewed the 'Malay problem' as a cultural or even racial characteristic, internal to the community and of very long standing. Again, it must be stressed that this was not the only trend in analysis. The *Mendaki Congress Papers* also give detailed accounts of specific problems and policies that affected the Malays in Singapore, and many of the individuals involved in Mendaki have produced critiques of broader structural features of Singapore society that increase the gap between the 'haves' and 'have-nots' of all races.[6] However, when the focus is on Malays as an ethnic group, the cultural and economic analyses tend to be divorced, and the supposed cultural deficiencies of Malay society are given priority. Mendaki has seen its main task to be that of reforming Malay attitudes and values.

In the months surrounding the launching of Mendaki, numerous items appeared in the press, most of which stressed the cultural aspect of the 'Malay problem'.[7] A review of a television documentary on Malay education concluded that the 'attitudes of the students and their parents towards education are largely responsible for the present situation'. In the review, the 'problem' was seen to be the number of Malays recorded in the national census as having no education or only primary education (*ST*, 29.5.82). While the low level of education among Malays was probably shocking to the reviewer and the literate, newspaper-reading public, the comparison was distorted by the failure to point out the high proportion of Chinese in the same predicament. An interview with some school principals about Malay education stated that the principals felt 'Malay students grow up in school with a host of built in social disadvantages. Just about everything–from their philosophy of life, to home environment–works against them and gives their non-Malay friends an edge over them.' The principals noted that Malay parents are generally uneducated, do not speak English, are often both working and that private tuition is too expensive for low-income families. Identical problems faced by Chinese students from low-income homes are not mentioned. Typically, the strongest cultural comments came not from a Chinese or Indian principal but from a Malay principal: 'Generally speaking we Malays have a different philosophy of life. We do not have the philosophy of wanting to be the best and the top most in the world... we are more content in seeking happiness through religion' (*ST*, 29.5.82).

The Mendaki programme to improve the standard of education and welfare in the Malay community moved beyond the criticism of the values of the 'old Malay' and concentrated on promoting the social values, based on Islam, that the leaders felt should characterize the new, successful, Malay. The focus of Mendaki's programme has been the family and through speeches, media coverage, booklets, marriage-guidance courses, sermons in the mosques, and community-level meetings, the effort has been made to instruct the Malay community on the techniques and values that will help them build a stable, happy, and successful Muslim family. A Mendaki booklet advises parents to work out a timetable for the children's study, television, and play time; to ensure that their children go home immediately after school and never go out after evening prayers; to restrict the use of television and telephone; to get to know their children's friends and ensure that they are suitable and studious; to make sure children pray regularly to instil discipline and knowledge of their religious obligations; to emphasize that Islam values education; to give their children religious greetings and blessings as they leave for school or before they go to bed, in order to engender a loving, concerned, and religious atmosphere in the home (Mendaki, n.d.). There are other programmes directed at newly-weds, giving advice on the proper roles and duties of husband and wife, domestic budgeting, and the care and nutrition of infants. The intention of Malay community leaders is to integrate young Muslim families in community-level programmes for child development, Islamic and secular kindergarten, primary school tuition, and so on, to guide them through the life cycle and produce a better educated and more Islamic 'quality' generation.[8]

The Malay leaders' effort to change Malay values assumes that, since poor Malay values were a source of Malay backwardness in the past, correct values together with certain techniques and organized support will enable Malays to change their lives and to succeed in future. There is an implicit recognition that not all Muslims can be successful, although this is seen to be more a result of the inadequacy of available financial and manpower resources to provide all Malays with the necessary preventative or remedial assistance, rather than an inevitable outcome of the structure of Singapore's educational and economic pyramid.

It was argued earlier that cultural values *do* shape the practices of daily life, and that these values are subject to constant change in the light of new ideas and new circumstances. It is possible that programmes such as Mendaki will become a source of new ideas that will be absorbed into the practices of the Malay community, heightening consciousness about education and the factors such as peer pressures, home study time, and nutrition that have an effect on educational performance. These are factors that are, to some extent, within the control of individuals and families and could be changed by motivated effort. The limitation of Mendaki's programme is that the formal public media are only one source of the cultural ideas that shape daily life, and they are heard, interpreted, and rejected or acted upon, with greater or lesser effectiveness as part of the continuous process of cultural and economic reproduction. The approach to social and economic change through cultural engineering lacks analysis and appreciation of the way in which the structure of opportunities both confronts individuals with absolute barriers–lack of money, failure in examinations, crowded home–and, in more indirect ways, shapes the entire cultural framework of day-to-day life.

An earlier chapter described the often subtle differences between the practices of upper- and lower-income Malays, and their relation to deep-seated Malay ideas and to unique economic, ethnic, and cultural conditions in Singapore. The likelihood is that programmes such as Mendaki will have greatest effect on the higher-income group with at least secondary education (about 20 per cent of the Malay population in the age group 20-30, Table 7.13) since these people are literate, and they are the ones most anxious to sustain their class position and to improve their children's chances in education. They are also the people who are most concerned about living and displaying an Islamic style of family and community life as the focus of their personal, class, and ethnic identity in Singapore. The limitation of the cultural-change approach is shown clearly in a project known as Restu, which preceded Mendaki. This programme for family reform states that the 'first step is a household where harmony and love coupled with comfortable earnings go hand in hand. Restu hopes to show parents how savings can produce the right home environment for study.' Savings will enable parents to work well, acquire training, and furnish the household with facilities for education, and 'financial stability will also lead to a desire on the part of parents to be ambitious for their children's education and general welfare' (*New Nation*, 5.9.81). It is undeniable that financial security does produce these effects, but given the low incomes of the majority of Malay households, it is doubtful whether exhortations to save, and a description of the operation of a comfortable, middle-class home, will really enable that ideal to be achieved. The underlying assumption is the recurrent orthodoxy that Malays are improvident, lacking in effort and ambition, and in need of persuasion to change their uniquely unprogressive cultural tendencies.

Ideology in Singapore:
Cultural Explanations for Structural Inequalities

'Ideology' was defined above as any aspect of a symbolic system which legitimizes domination through becoming integrated as part of the cultural knowledge of practical daily life or through its acceptance in more formalized versions as an orthodoxy. It was stated that this ideological effect is not necessarily planned or intended, nor is it identical for all sections of the population for whom cultural and economic circumstances differ. Here it will be argued that the idea that Malays are behind the other races economically because of factors internal to Malay culture, has come to play an ideological role in legitimizing the inequalities in educational opportunity and in economic reward that have characterized Singapore since Independence.

The idea that the backwardness of Malay society stems from faults deep within itself is prominent and widespread as indicated above. The idea has dominated the thinking, the politics, and the programmes of the Malay élite since Independence. Since many among this élite are themselves examples of how *some* individuals from humble origins can, with ability and motivation, overcome their disadvantages and achieve mobility through a 'meritocratic' system, they are inclined to believe that mobility is possible for everyone. In interviews, many stated that their personal philosophy of life and motivation to succeed is different from that of the mass of Malays whom they believe to be impeded by cultural values from achieving success as individuals or as a community. At various times the Malay leadership has voiced its fears that those who begin from a disadvantaged position in a 'meritocratic' order may not be able to 'catch up', but this has been taken to mean that the Malays need additional help of some kind to make up for historical factors that once held them back. It is not generally perceived that the inability of the lower socio-economic stratum to achieve mobility is a perennial rather than occasional or temporary feature of a 'meritocratic order', or one that is unique to Malays. Where the problem of the general impermeability of the national education and economic system is perceived, there is a tendency for the Malay leadership to see it as outside its own domain of control or responsibility, and their focus on affairs internal to the Malay community has helped to accommodate them to the prevailing economic order.

The view of the Malay predicament from the lower end of the economic hierarchy is rather different. The idea that Malays have cultural impediments to progress has been incorporated into the practices of daily life through the common idea that the pattern of interaction in lower-income Malay neighbourhoods is undesirable, and that lower-class Malay society has to be avoided if mobility is to be achieved. There is also the idea that discretion and studiousness are the characteristic cultural traits of the Chinese race, though they are also the characteristics associated with higher-income and higher-status Malays. Chapter 8 described the complexity of the way ideas of class, ethnicity, and culture have become superimposed upon one another. This is a factor that helps to legitimize the current distribution of opportunities and rewards in Singapore.

In contradistinction to their generalized statements about the cultural characteristics and economic potential of Singapore's main ethnic groups, it is noteworthy that individual Malays recounting their own personal life stories never mention the idea that Malays are culturally deficient as a reason why they themselves ended up with low education or low incomes. They attribute their individual fate to personal factors, such as a dislike of school, an inability to concentrate in school, the influence of peers, and the desire to work, or to circumstantial factors such as the death or divorce of a parent, or lack of money. All of these factors are also mentioned by Chinese youths as reasons for dropping out of school (Leong Choon Cheong, 1978). There is considerable faith among the Singapore Malay populace in the openness of the education system, despite the difficulties that confront the majority of the nation's population in their attempt to achieve social mobility through educational means.

For Malays in both higher- and lower-income groups, personal experiences of racial discrimination in working life form the basis of the most common generalization about why Malays have not progressed in Singapore. The widespread and deeply held belief among Malays in Singapore is that their problems and disadvantages have been imposed on them on a racial basis by the Chinese majority. Discrimination by Chinese against Malays is based on the Chinese opinion that Malays are culturally inferior and incapable of hard work, and this view has probably been reinforced and confirmed by the pervasive orthodoxy about Malay cultural inadequacies that Chinese read in the press. There is no doubt that some discrimination does occur, but it is also the case that the tendency to see Singapore society as divided on racial lines, and Malay disadvantages as stemming from race, has distracted Malay attention from the process of economic differentiation, which also has a great impact in shaping the conditions of their daily lives.

Although the Malay cultural-weakness orthodoxy has been purveyed mostly by the Malay elite themselves, there are clear signs that Chinese in the government hold similar views, both about Malay weaknesses and about Chinese strengths. Singapore's economic development since 1819 is usually portrayed in official publications, such as tourist brochures, school textbooks, and commemorative histories, as the result of the industry and often also the *intelligence* of the Chinese migrants (*Singapore: A Salute to Singapore*, 1984: 19). Members of the government have expressed the view that the best of Chinese culture is found in Confucian moral concepts expressed in the Chinese language, and instilled through the discipline of a traditional Chinese education. Official efforts have been made to preserve Chinese institutions from the supposed degeneracy of Western culture. Paradoxically, at the same time as government concern is expressed over the possible loss or erosion of Chinese culture, there exists a strong official and popular tendency in Singapore to see culture and race as integrally linked (Clammer, 1981: 224, 1982: 130), a view which permits positive Chinese cultural characteristics to be considered permanently inherent in the Chinese race.

Whether it is seen in terms of race, culture, or specific cultural codes such as Confucianism, Chineseness is invariably associated in public discourse with diligence, material success, and progress, as well as other virtues such as filial piety.[9] For example, the Deputy Prime Minister, Goh Keng Swee, attributed the outstanding economic success of Taiwan, South Korea, and Singapore to their Confucian ethics, implying that the other races present in Singapore have made less significant contributions (*ST*, 4.2.82). Policies introduced in the early 1980s to encourage the importation of workers from these 'Confucian' areas instead of workers from the Indian subcontinent or elsewhere in South-East Asia, and the attempt to recruit Chinese professionals and entrepreneurs from around the world to work and preferably settle in Singapore (*FEER*, 6.10.83: 29), are based on the government's conviction that the Chinese are superior in work ethics, industry, and intelligence.

A statement which shows the way in which reified cultural traits attributed to racial groups are used to explain the distribution of economic resources and rewards in Singapore was made by the Deputy Prime Minister (Goh, 1977: 45):

In the ethos of Malay society, the unrelenting pursuit of an objective, like the accumulation of wealth, is not held in esteem. In fact it is condemned as inconsistent with gracious living on which they place much importance. The spectacle of the Chinaman working like a demon possessed and ruthlessly brushing aside any one or any obstacle that stands in his way is not one that arouses Malay admiration. And yet without this passion for wealth generating a fanatical determination to accumulate, is it possible for the Malays to achieve the economic success of the Chinese?

Goh attributes a 'fanatical determination to accumulate' to an entire racial group, but this characteristic is not prominent in the life histories of Singapore Chinese youths collected by Leong Choon Cheong (1978), nor is 'success' the characteristic of 40 per cent of the Chinese male working population with incomes below $400 in 1980 (Table 7.4), nor is a preoccupation with gracious living evident among the Malays encountered in the course of this research. It *is* true that there are differences in the way Malays and Chinese handle economic and interpersonal relations, and these differences do have some effect on their relative economic performance. Some of the major differences have been investigated here in considerable depth. But the abstraction, reification, and generalization of cultural traits evident in statements such as that by Goh above, distorts rather than contributes to an understanding of both the cultural and economic development of Singapore.

Nagata (1975: 124) argues that in Malaysia, the rich Chinese generally fail to perceive the existence of the Chinese poor, so strongly is their vision shaped by the image of the industrious and successful Chinese and the poor, lazy Malay. In Singapore, too, the Chinese élite holds to the view of Chinese cultural superiority and Malay inferiority, and the distortion introduced by these racial images may partly account for the lack of attention paid to the predicament of the poorly educated, low-income Chinese who form the majority of Singapore's population (Cheah, 1977: 141). The supposition that Malays are deficient in diligence and work ethic affects the material opportunities of Malays through the practice of discrimination. It also affects the way economic differentiation–which now coincides in part with a widening gap between Chinese and Malays–is perceived, managed, and justified by the Chinese economic and political élite of Singapore.

In the Chinese lower-income group, the image of the backward and lazy Malay is strongly entrenched, and it is possible that this gives even the poorest Chinese a certain sense of superiority. However, the Malays are a minority in Singapore, and do not figure prominently in the daily lives of most Chinese. Leong Choon Cheong's (1978) study of Chinese working-class youths indicates the predominance of a model of rich and poor, powerful and powerless, in their world-view, and the racial image, though present, probably does not play a major part in legitimizing this state of affairs.

Abercrombie and Turner (1978) argue that ideology plays its most important role in justifying the advantages enjoyed by the dominant groups of society, to *themselves*. It is less important to the masses who are often illiterate and less exposed to the prevailing orthodoxies, and who are affected more by the 'dull compulsion' that requires them to work to sustain themselves than by a positive legitimization of the division of power and rewards. This appears to be true of the Malay cultural-weakness orthodoxy in Singapore, which reassures the Malay élite that they have earned their merit by perceiving and criticizing the cultural traits that restrain the progress of the Malay masses, and reassures the Chinese élite that their success is rightfully theirs by virtue of cultural and racial superiority. In the arena of public discourse, and in the arena of daily life for all Malays, the prominence of ideas about ethnicity distracts attention from the process of economic differentiation taking place in Singapore and therefore plays a part in explaining and legitimizing the unequal distribution of power and rewards.

Clammer has noted the political implications of the model of race and culture espoused by the Singapore government, and writes (1982: 134) that 'individuals or communities can be organized or mobilized on the basis of race, can be subdivided from one another on the same basis, and can be played off against one another if need be'. He argues (1982: 135) that the racial model of society 'provides a kind of structural principle in the light of which, within Singapore, other problems can be interpreted and all can be simplified'. He continues (1982: 138):

Ethnicity in Singapore structures, but it also distracts from the profound underlying economic developments taking place in the republic. Indeed a key feature of the Singapore ideology is to try to show that while the economy is dynamic, ethnicity is static.... [E]thnicity has certainly succeeded in upstaging class in the way in which social stratification is believed to be organized–a view which has the advantage that ethnicity, unlike class, is seen as being politically and economically neutral.

Clammer (1982: 136) uses the term 'ideology' to mean 'a model of reality' which is the sense in which the present work has used the term 'culture'. The analysis of the Malay community given here supports his contention that ideas about ethnicity have become an important part of the cultural framework of Singapore life.

Chua's (1983b) discussion of ideology in Singapore uses the term to refer to the legitimization of sectional interests, in a way similar to the usage adopted here. Chua argues that ideology is actively propagated and inculcated by the government, although once an ideological system of ideas is successfully established as a hegemony, it becomes invisible as ideology and appears to be merely practical, natural, common sense.

Contrary to a purposive or conspiratorial view of ideology, this work has tried to show the complex and diverse origins of the set of beliefs about culture and ethnicity, and about particular cultures and ethnic groups, that have come to play an ideological role in Singapore. It has argued that differences in the cultural framework within which Malays and Chinese organize their economic lives, especially with regard to entrepreneurship, have put Malays at an economic disadvantage in Singapore since 1959, and supported an idea that Malays are culturally inferior which, in turn, has been a source of discrimination against them. Practices within the household and community of the lower-income strata of each ethnic group tend to cause the economic disadvantage of one generation to be passed on to the next, despite the apparent opportunities of an 'open' economic and education system. None of this is planned, intended, or imposed by the political and economic élite of Singapore. The reproduction of culture, ethnicity, class, and ideology all takes place as the direct or indirect consequence of the practices of day-to-day life which are complex, creative, and not amenable to social engineering or control, although government policies on matters such as housing, education, and pensions do establish some of the practical conditions and parameters. Governments, like ordinary citizens, contribute to the making of history, but *not* always in circumstances that they choose or fully control.

1. The tendency to reify culture, and then to use it to explain economic superiority and inferiority, has been the main weakness of studies of the 'culture of poverty' (Lewis, 1969), studies of the working class in Britain, and studies of the economic performance of ethnic groups. See the critiques by Steinberg (1981) and Brook and Finn (1978). See Gans (1969) for an example of a reified analysis of class culture. See Smith (1975) for an account of rapid cultural change in improved economic circumstances.

2. Thanks are due to Haji Sha'ari Tadin for the loan of PPPPM and Gerakan Obor archive material which permitted the author some familiarization with Malay social analysis of the 1960s.

3. See Dzulfiqhar Mohd. M. (1979) and *Insaf* (4.7, 5.1, 7.3, 8.2, 10.2) for further ndergraduate writing on Malay values, Islam, and the 'Malay problem'. See Nagata (1980-1) for a discussion of the Islamic revival in Malaysia and the use of Islam by the Malay political and economic élite as a model for social change. See Gellner (1981) for a more general discussion of the modernizing potential of Islam.

4. There are numerous references to the supposed rural orientation of Singapore Malays in academic literature. See Yang Razalie Kassim (1979: 125); Pang Eng Fong (1983: 332); Nagata (1979: 84); Hanna (1966: 2-6).

5. The same quotation was used by Yang Razalie Kassim (1979), and in the *Far Eastern Economic Review* feature on Singapore Malays (*FEER*, 28.6.84), each user regarding the previous user as an 'authority' and thereby helping to develop and confirm the orthodox interpretation of the 'Malay problem'.

6. See articles by Malay university students on the education system and Eugenics programmes (*Insaf*, 9.2, 9.3, 11.2); Ma'arof Salleh's critique of the graduate mothers policy in *Muzakarah* 1 (1984), and Mansor Haji Sukeimi (1979).

7. The press coverage of Mendaki carried headlines associating Malays with problems and the need to change attitudes and make efforts. This was intended to awaken the Malay public to their faults, but it must have inadvertently confirmed the stereotypes of Malays held by the non-Malay public. The reader is referred to a selection of headlines listed in the bibliography under *Straits Times*, especially during the period January to November 1982, and *ST* (29.9.84).

8. See the Mendaki concept paper 'Keluarga Bahagia' (1982) for the programme for a new Muslim family; the mosque sermon 'Jurang Generasi', *Warita* 20 (1983) which stresses that broken families have caused social problems such as drug addiction while happy families produce successful children; the 1984 marriage preparation course by the religious organization PERDAUS, and the lecture 'Perkahwinan dan Undang menurut Agama Islam' by Ustaz Haji Abu Baker Hashim on 14.10.84 at the Queenstown Mosque.

9. See Gopinathan (1979: 283, 288-9) for a further discussion of this issue.

12 Conclusion

THIS work has explored the relationship between the form of cultural practices in the Malay household and community, and the economic structure of Singapore. It has attempted a *historical* analysis in which neither culture and forms of life nor material conditions would be treated separately nor regarded as static and pre-given. Analysis of the minutiae of relations within the household, combined with national statistical data and comparative ethnography, has been used to show how the practices of daily life in the Malay household are *structured* by economic conditions, though not determined by them. In turn, culturally informed day-to-day practices are also *structuring*, though not in a direct and voluntaristic way, as they have both immediate and longer-term, unintended effects on the position of Malay households in the national economy and on the wider processes of ethnic and economic differentiation in Singapore.

An idea central to the cultural form of the Malay household is that of the economic integrity of the individual. All social and economic relations with others are seen as personally constructed on the basis of individual will, and they are created, sustained, and demonstrated by means of material exchanges. It was found that the idea of the gift dominates the pattern of relations in the conjugal and filial bond as well as between individuals outside the household. The positive value given to kinship sentiment, and the emphasis on parental, marital, and filial responsibility, are expressions of the ideal state of relationships. However, it is recognized that this ideal does not exist as an inevitable outcome of kinship, nor is it imposed by law, but it is an ideal that can be accomplished by individuals who are willing to enter into relationships and exchanges which reduce their personal economic autonomy but enhance their social- and self-esteem. The immediate economic effect of this cultural arrangement is that there is no economic corporateness and only limited co-operation within the elementary family unit, and limited incentive to accumulate more wealth than the individual requires for his own lifetime.

There have been deep changes in the relationships between husband and wife, and parent and child, in the economic context of Singapore's urban wage economy. These changes are related to the wage-earning power of the various individual members of the family, and the change in the type of goods which the older generation gives to the young. The impact of these structural features of the Singapore economy on the Malay household was analysed by means of comparisons with the rural Malay and Javanese hinterland. Although there are continuities in some of the cultural ideas that characterize rural and urban Malay society, the contemporary form of the Malay household in Singapore cannot be seen as the mere transference of patterns which are static or pre-given. Description of the details of the negotiations and transactions within the household showed that the contemporary Malay household pattern in Singapore is a pattern that is in the process of being actively and creatively produced.

The pattern of Malay individualism described here is similar to that found in Britain historically, which now characterizes much of the Western developed world (Macfarlane, 1986). Its features are bilateral kinship, an absence of stress on descent, individual ownership of property, and the predominance of the elementary family as the unit of co-residence. The Western kinship system has been argued by social historians to have had an important role in the early development of Western capitalism (Macfarlane, 1978 and 1986; Caldwell, 1976). Ironically, in the Malay case, this family pattern has not been conducive to economic development or to entrepreneurship. It was argued that Malay society gives great emphasis to the idea of kinship sentiment, and this tempers the practical expression of individualism. Individuals are recognized to have full rights to their own labour power and property, but if the individual wishes to sustain the social element in interpersonal relationships, these rights should not be openly or aggressively asserted. A reluctance to commercialize family relationships and a tendency to personalize commercial relationships inhibit the development of contractual relations, explicit economic calculation, and the possibility and acceptability of personal advancement through entrepreneurship. The numerical and commercial dominance of the Chinese in Singapore has further narrowed the range of entrepreneurial opportunities for Malays. The Chinese presence has deeply affected Malay ideas about their own culture and ethnic identity, and has influenced the way Malays conduct economic and social relations among themselves.

Practices within the Malay and Chinese household, historical conditions, the process of economic differentiation, and some policies pursued by the government, have contributed to the emergence of a situation in Singapore where ethnic groups have become associated with economic performance. This association forms part of the cultural framework of daily life and public discourse in Singapore, and it plays an ideological role in the legitimization of inequality. It was shown that the practices of day-to-day life contribute in unintended ways to building, sustaining, and legitimizing the economic conditions of society which then appear to particular historically located actors as pre-given, objective, and inevitable structures of opportunity and constraint.

The analysis presented here has ranged from detailed ethnography to discussion of broader comparative and theoretical themes. It links micro-

and macro-data not by merely juxtaposing one level to another, but by seeking to understand the complex ways in which these levels interact and mutually shape one another. Ethnography is enriched by a historical and comparative approach which provides a perspective on time and space, and permits the interpretation of social life as a continuous process of change. Examination of quantitative trends and broader economic, ethnic, and ideological systems reveals some of the cumulative consequences of micro-level practices while, at the same time, the macro phenomena themselves can be better interpreted by reference to a meaningful, cultural dimension.

Abstract, ahistorical observations on cultural and economic trends have been a source of some significant misapprehensions about the Singapore Malay community. Through the data and analytical framework presented in this book, and the discussions and further research it may generate, a fuller understanding of the position of Malays in Singapore may be achieved.

Glossary

abangan	people adhering to the indigenous traditions of Java
adat	custom, in the widest sense
bangkit	to raise old matters, used in reference to debts generated by the receipt of kindness or favour
bantu	help, assistance
bela	to bring up or support a child (or other relative)
belanja kahwin	money given by the groom to the bride upon marriage
berkira	calculating, mean, ungenerous
berumahtangga	to marry and set up home
busuk hati	evil
darah	blood; relationship defined by blood ties
duit	money
duit penat	compensation for expenditure of effort
ganja	marijuana
garang	stern, fierce, quick-tempered
gotong-royong	co-operation, helping one another
gurau	sportive, liking to joke and laugh
hadiah	a gift
haj	Muslim pilgrimage to Mecca
harta pesaka	ancestral property, inherited wealth
harta sepencarian	the joint earnings of husband and wife
harta syarikat	the joint earnings of husband and wife; a partnership; a mercantile firm
hutang	debt
hutang budi	a debt of kindness
iri hati	jealous
jaga	to watch, guard, or care for a child (or other relative)
jodoh	a match, a pair; counterpart, twin soul
kampong	a group of houses, village or hamlet in either a rural or urban area
kasi	to give
kasih	love and affection
kasihan	kindness, favour, pity, an unfortunate thing; also related to *kasi*
kenduri	ritual feast with prayers to mark an occasion of personal or general religious significance
Kenduri Arwah	prayers for a deceased, held at specific intervals after the death

Kenduri Ruahan	prayers for all deceased Muslims or 'All Souls Day' held in the Muslim month of Syaaban
kesayangan	love, kinship sentiment
kongsi	a Chinese association; a group of Chinese workers under a contractor; Chinese word used in the Singapore Malay language to mean sharing or partnership
manja	indulge or spoil with affection
muafakat	agreement or settlement by discussion
muta'ah	consolatory gift that can be awarded to an aggrieved party upon divorce
nafkah	means of livelihood; a living; alimony; part of the marital contract
orang baik	a good and worthy person
orang tebusan	bonded labourer, held by a debt to the employer
pahala	personal religious grace; merit
panas hati	angry
pelihara	to cherish, bring up, protect, or nurture a child (or other relative)
Peranakan	early Chinese settlers to Malaysia or Singapore who developed a distinctive culture incorporating some Malay elements
pondok	multifamily dwelling once common in the Singapore Bawean community
pusing	revolve, rotate, circulate
rezeki	livelihood; source of income; economic destiny
sakit hati	upset, resentful
sakit mata	envious
santri	Javanese adhering to orthodox Islamic practices
sombong	proud, arrogrant
susah hati	sorrowful
Syariah Court	Muslim court which, in Singapore, deals mainly with family law matters such as divorce and inheritance
tanggungjawab	responsibility
tidak puas hati	dissatisfied
tidak sedap	unpleasant
tolong	help, assist
tumpang	to 'take advantage' in the sense of hitching a ride
tumpang hartanya	to 'take advantage' of another's wealth or property, for example by living in a house inherited from an ancestor
untung	luck, good fortune; profit; fate

Bibliography

Abdullah bin M. Baginda (1967), 'Our Baweanese People', *Intisari*, 2: 16-60.
Abercrombie, Nicholas and Turner, Bryan S. (1978), 'The Dominant Ideology Thesis', *British Journal of Sociology*, 29(2): 149-70.
Abu Bakar Hashim, Ustaz, Haji (1984), 'Perkahwinan dan Undang Undang Menurut Agama Islam', lecture given at the Queenstown Mosque, 14 October 1984.
Ahmad Ibrahim (1978), *Family Law in Malaysia and Singapore*, Singapore, Malayan Law Journal.
Ahmad Mattar (1979), 'The Singapore Malays: Their Education and Role in National Development', in *PAP 25th Anniversary Issue 1954-1979*, pp. 78-85.
Alatas, Syed Hussein (1972), *Modernization and Social Change*, Sydney, Angus & Robertson.
_____ (1977), *The Myth of the Lazy Native*, London, Frank Cass.
Alexander, J. and Alexander, P. (1982), 'Shared Poverty as Ideology: Agrarian Relations in Colonial Java', *Man*, 17: 597-619.
Aljunied, S. Zahra (1980), 'Ethnic Distribution of Employment in Singapore: The Malays', academic exercise in Economics and Statistics, National University of Singapore.
Bailey, Connor (1983), *The Sociology of Production in Rural Malay Society*, Kuala Lumpur, Oxford University Press.
Banks, David J. (1972), 'Changing Kinship in North Malaysia', *American Anthropologist*, 74: 1254-75.
_____ (1976), 'Islam and Inheritance in Malaya', *American Ethnologist*, 3: 573-86.
_____ (1983), *Malay Kinship*, Philadelphia, Institute for the Study of Human Resources.
Barth, Frederick (1981), 'Ethnic Groups and Boundaries', in *Process and Form in Social Life*, London, Routledge & Kegan Paul, 1: 198-227.
Bedlington, Stanley S. (1971), 'Malays of Singapore: Values in Conflict?', *Sedar*, 3: 43-55.
_____ (1974), 'The Singapore Malay Community: The Politics of State Integration', Ph.D. dissertation in Political Science, Cornell University.
Benjamin, Geoffrey (1976), 'The Cultural Logic of Singapore's Multiracialism', in Riaz Hassan (ed.), *Singapore: Society in Transition*, Kuala Lumpur, Oxford University Press, pp. 115-33.
Betts, Russell Henry (1975), 'Multiracialism, Meritocracy and the Malays of Singapore', Ph.D. dissertation in Political Science, Massachusetts Institute of Technology.
Bourdieu, Pierre (1977), *Outline of a Theory of Practice*, Cambridge, Cambridge University Press.

Brook, Eve and Finn, Dann (1978), 'Working Class Images of Society and Community Studies', in Centre for Contemporary Cultural Studies, *On Ideology*, London, Hutchinson, pp. 125-43.

Caldwell, J. C. (1976), 'Towards a Restatement of Demographic Transition Theory', *Population and Development Review*, 2: 321-66.

Carstens, Sharon (1975), *Chinese Associations in Singapore Society*, Occasional Paper No. 37, Singapore, Institute for Southeast Asian Studies.

Chang Chen Tung, Ong Jin Hui, and Chen, Peter S. J. (1980), 'Culture and Fertility: The Case of Singapore', Singapore, Institute for Southeast Asian Studies, Research Notes and Discussion Papers, 21.

Chang Soo (1960), 'Chinese Lineage Settlements in Singapore', Diploma in Social Studies, University of Malaya.

Chayanov, A. V. (1966), *On The Theory of Peasant Economy*, trans. D. Thorner, B. Kerblay, and R. E. F. Smith, Homewood, Illinois, American Economic Association.

Cheah Hock Beng (1977), 'A Study of Poverty in Singapore', Masters dissertation in Sociology, University of Singapore.

Chee Heng Leng and Chan Chee Khoon (1984), *Designer Genes: IQ, Ideology and Biology*, Kuala Lumpur, INSAN.

Chen, Peter S. J., Kuo, Eddie C. Y., and Chung, Betty Jamie (1982), *The Dilemma of Parenthood: A Study of the Value of Children in Singapore*, Singapore, Maruzen Asia.

Cheng Siok Hwa (1980), 'Recent Trends in Female Labour Force Participation in Singapore', *Southeast Asian Journal of Social Science*, 8(1-2): 20-39.

Chua Beng Huat (1983a), 'Some Methodological Issues in Resettlement Study in Singapore', paper presented at International Ethnographic Association Meeting, Vancouver.

_____ (1983b), 'Reopening Ideological Discussion in Singapore: A New Theoretical Direction', *Southeast Asian Journal of Social Science*, 11(2): 31-45.

Chung, Betty Jamie, Chen, Peter S. J., Kuo, Eddie C. Y. and Nirmala Purushotam (1981), *The Dynamics of Child Rearing Decisions: The Singapore Experience*, Singapore, Maruzen Asia.

Clammer, J. (1981), 'Modernization and Cultural Values: The Paradoxes of Transition in Singapore', in R. E. Vente, K. S. Bhatal, and R. M. Nakhooda (eds.), *Cultural Heritage v. Technological Development*, Singapore, Maruzen Asia, pp. 223-40.

_____ (1982), 'The Institutionalization of Ethnicity: The Culture of Ethnicity in Singapore', *Ethnic and Racial Studies*, 5(2): 127-39.

_____ (1983), 'Chinese Ethnicity and Political Culture in Singapore', in Linda Y. C. Lim and Peter Gosling (eds.), *The Chinese in Southeast Asia*, Singapore, Maruzen Asia, Vol. 2, pp. 266-84.

Cohen, Myron L. (1970), 'Developmental Processes in the Chinese Domestic Group', in Maurice Freedman (ed.), *Family and Kinship in Chinese Society*, Stanford, Stanford University Press, pp. 21-36.

Coope, A. E. (1976), *Macmillans Malay-English Dictionary*, London, Macmillan.

Dewey, Alice G. (1962), *Peasant Marketing in Java*, New York, Free Press of Glencoe.

_____ (1964), 'Capital, Credit and Saving in Javanese Marketing', in Raymond Firth and B. S. Yamey (eds.), *Capital, Saving and Credit in Peasant Societies*, London, George Allen & Unwin, pp. 230-55.

Deyo, Frederick C. (1983), 'Chinese Management Practices and Work Commitment in Comparative Perspective', in Linda Y. C. Lim and Peter Gosling (eds.), *The Chinese in Southeast Asia*, Singapore, Maruzen Asia, Vol. 2, pp. 215-29.

Djamour, Judith (1959), *Malay Kinship and Marriage in Singapore*, LSE Monographs on Social Anthropology, 21, London, Athlone Press.

_____(1966), *The Muslim Matrimonial Court in Singapore*, London, Athlone Press.

Dzulfiqhar Mohd. M. (1979), 'Dakwah Movement in Singapore: Its Social and Political Orientations', academic exercise in Political Science, University of Singapore.

Freedman, Maurice (1957), *Chinese Family and Marriage in Singapore*, London, HMSO.

_____(1970), *Family and Kinship in Chinese Society*, Stanford, Stanford University Press.

_____(1979), *The Study of Chinese Society, Selected essays by Maurice Freedman with Introduction by G. William Skinner*, Stanford, Stanford University Press.

Gans, Herbert (1969), 'Class Subcultures in American Society', in Heller (ed.), *Structured Social Inequality*, London, Collier Macmillan, pp. 270-6.

Geertz, Clifford (1956), 'Religious Belief and Economic Behaviour in a Central Javanese Town: Some Preliminary Considerations', *Economic Development and Cultural Change*, 4: 134-58.

_____(1960), *The Religion of Java*, New York, Free Press of Glencoe.

_____(1963), *Peddlers and Princes: Social Development and Economic Change in Two Indonesian Towns*, Chicago, University of Chicago Press.

_____(1973), *The Interpretation of Culture*, New York, Basic Books.

Geertz, Hildred (1961), *The Javanese Family*, New York, Free Press of Glencoe.

Gellner, Ernest (1981), *Muslim Society*, Cambridge, Cambridge University Press.

Giddens, Anthony (1979), *Central Problems in Social Theory*, London, Macmillan.

_____(1981), *A Contemporary Critique of Historical Materialism*, London, Macmillan.

_____(1985), 'Marx's Correct Views on Everything', *Theory and Society*, 14: 167-74.

Goh Keng Swee (1958), *Urban Incomes and Housing: A Report on the Social Survey of Singapore 1953-1954*, Singapore, Government Printer.

_____(1977), *The Practices of Economic Growth*, Singapore, Federal Publications.

_____(1978), *Report on the Ministry of Education*, Singapore, Government Printers Office.

_____(1980), 'Equality in Education', *Speeches*, 3(12): 30-3.

Gopinathan, S. (1974), *Towards a National Education System in Singapore 1945-1973*, Singapore, Oxford University Press.

_____(1979), 'Singapore's Language Policies: Strategies for a Plural Society', in *Southeast Asian Affairs*, Singapore, Institute for Southeast Asian Studies & Heinemann Educational Books (Asia), pp. 280-95.

Gordon, Shirle (1965), 'Marriage and Divorce in the Eleven States of Malaya and Singapore', *Intisari*, 2(2): 23-32.

_____(1968-70), 'Malay Education, Analysis and Recommendations', *Intisari*, 3(3): 35-50.

Gordon, Shirle, and Gunn, G. T. (1968-70), 'Malay Secondary School Students—Survey Findings', *Intisari*, 3(3): 52-72.

Hanna, W. A. (1966), *The Malay's Singapore*, American Universities Field Staff

Reports Southeast Asia Series, 14(2-6).

Hardjono, Joan (1987), *Land, Labour and Livelihood in a West Java Village*, Yogyakarta, Gadjah Mada University Press.

Harevin, Tamara (1977), 'Family Time and Historical Time', *Daedalus*, 106(2): 57-70.

Harris, Olivia (1981), 'Households as Natural Units', in Kate Young, Carol Wolkowitz, and Roslyn McCullagh (eds.), *Of Marriage and the Market: Women's Subordination in International Perspective*, London, CSE Books, pp. 49-68.

Harris, Olivia and Young, Kate (1981), 'Engendered Structures: Some Problems in the Analysis of Reproduction', in Joel S. Kahn and Joseph R. Llobera (eds.), *The Anthropology of Pre-Capitalist Societies*, London, Macmillan, pp. 109-47.

Hart, Gillian (1986), *Power, Labor, and Livelihood: Processes of Change in Rural Java*, Berkeley, University of California Press.

Hartman, Heidi (1981), 'The Family as the Locus of Gender, Class and Political Struggle: The Example of Housework', *Signs*, 6(3): 366-94.

Harun Aminurrashid (1932), *Melor Kuala Lumpur*, 2nd ed., Singapore, 1962.

Household Research Working Group (1982), 'Household Structures and the World Economy', paper presented at a joint seminar 'Households and the World Economy', Fernand Braudel Centre, State University of New York, Binghampton, and Sociology of Development Research Centre, University of Bielefeld, Bielefeld.

Ismail Kassim (1974), *Problems of Elite Cohesion: Perspectives from a Minority Community*, Singapore, Singapore University Press.

Jay, Robert (1969), *Javanese Villagers: Social Relations in Rural Modjokuto*, Cambridge, Mass., Massachusetts Institute of Technology.

Jenkins, Richard (1983), *Lads, Citizens and Ordinary Kids: Working-class Youth Life-styles in Belfast*, London, Routledge & Kegan Paul.

'Jurang Generasi' (1983), *Warita*, 20: 4-5.

Kamal Salih *et al.* (1983), 'Industrialization, Urbanization and Labour Force Formation in Penang: A Research Proposal', paper presented at Research Seminar on Third World Urbanization and the Household Economy, University Sains Malaysia, Penang.

Koentjaraningrat, R. M. (1961), *Some Social Anthropological Observations on Gotong Royong Practices in Two Villages of Central Java*, Ithaca, Modern Indonesia Project, Cornell University.

_____ (1962), 'Review of *The Javanese Family*, by Hildred Geertz', *American Anthropologist*, 64: 872-4.

Lai, Alice (1973), 'School Under-achievers in an Urban Slum in Singapore', academic exercise in Social Work, University of Singapore.

Lau Hong Thye (1974), 'The Social Structure of Small Chinese Business Firms in Singapore', academic exercise in Sociology, University of Singapore.

Lee Kuan Yew (1980), 'Prime Minister's Opening Speech in Malay at the National Day Rally on 19 Aug. 1980', *Lee Kuan Yew's Speeches*.

_____ (1982a), 'Three Generation Family is an Asset to the Nation', *Speeches*, 5(8): 17-19.

_____ (1982b), 'Mendaki's Task is to Raise the Education of all Malays', *Speeches*, 5(2): 5-25.

_____ (1983), 'Speech by Prime Minister Lee Kuan Yew at the National Day Cultural Show and Rally on 14 Aug. 1983', *Lee Kuan Yew's Speeches*.

Leong Choon Cheong (1978), *Youth in the Army*, Singapore, Federal Publications.

Leong Pui Ling (1982), 'The Modified Extended Family in Singapore: An Exploratory Analysis', academic exercise in Sociology, National University of Singapore.
Lewis, Oscar (1969), 'Culture and Poverty: Critique and Counterproposals', in Charles A. Valentine (ed.), *Current Anthropology*, 10(2-3): 181-201.
Liebow, E. (1967), *Tally's Corner*, Boston, Little Brown.
Lim Guek Poh (1974), 'Factory Girls in Jurong', academic exercise in Sociology, University of Singapore.
Lim Kim Huay (1960), 'The Supply of Labour to the Building and Construction Industry in Singapore', academic exercise in Economics, University of Malaya (Singapore).
Lim, Linda Y. C. (1983a), 'Chinese Economic Activity in Southeast Asia: An Introductory Review', in Linda Y. C. Lim and Peter Gosling (eds.), *The Chinese in Southeast Asia*, Singapore, Maruzen Asia, Vol. 1, pp. 1-29.
———(1983b), 'Chinese Business, Multinationals and the State: Manufacturing for Export in Malaysia and Singapore', in Linda Y. C. Lim and Peter Gosling (eds.), *The Chinese in Southeast Asia*, Singapore, Maruzen Asia, Vol. 1, pp. 245-73.
Ma'arof Salleh (1984), *Muzakarah*, 1.
Macfarlane, Alan (1978), 'Modes of Reproduction', in Geoffrey Hawthorne (ed.), *Population and Development*, London, Frank Cass, pp. 100-20.
———(1986), *Love and Marriage in England 1300-1840*, Oxford, Blackwell.
McGee, T. G. (1973), 'Peasants in the Cities: A Paradox, A Paradox, A Most Ingenious Paradox', *Human Organization*, 32(2): 135-42.
———(1982), 'Women Workers or Working Women?', paper presented at the Conference on Women in the Urban and Industrial Workforce: Southeast and East Asia, Manila.
McGlothlin, W. H. (1980), 'The Singapore Heroin Control Programme', *Bulletin on Narcotics*, 32(1): 1-14.
McKinley, Robert H. (1975), 'A Knife Cutting Water: Child Transfers and Siblingship among Urban Malays', Ph.D. dissertation in Anthropology, University of Michigan.
McLellan, Susan (1985), 'Reciprocity or Exploitation? Mothers and Daughters in the Changing Economy of Rural Malaysia', Working Paper on Women in International Development No. 93, Michigan State University.
Madge, Charles (1974), 'The Relevance of Family Patterns in the Process of Modernization in East Asia', in Robert J. Smith (ed.), *Social Organization and the Applications of Anthropology*, Ithaca, Cornell University Press, pp. 161-95.
Mahathir bin Mohamed (1970), *The Malay Dilemma*, Singapore, Times Books International.
Mak Lau Fong (1973), 'The Forgotten and Rejected Community: A Sociological Study of Chinese Secret Societies in Singapore and West Malaysia', Sociology Working Paper 18, University of Singapore.
Malay Youth Literary Association (PPPPM), (1965), *Penyata Tahun 1965*.
Mansor Haji Sukeimi (1979), *Dear Dr Goh*, annex to debate on the *Report on the Ministry of Education*, Parliament of Singapore, 29 March 1979 and pamphlet.
Marx, Karl (1951), 'The Eighteenth Brumaire of Louis Bonaparte', in *Marx-Engels Selected Works*, London, Lawrence & Wishart, Vol. 1.
Matheson, Virginia (1979), 'Concepts of Malay Ethos in Indigenous Malay Writings', *Journal of Southeast Asian Studies*, 19(2): 351-71.
Mauss, Marcel (1966), *The Gift*, trans. Ian Cunnison, London, Cohen & West.
Maxwell, Hon. W. E. (1884), 'The Law and Custom of the Malays with Refer-

ence to the Tenure of Land', *Journal of the Straits Branch of the Royal Asiatic Society*, 13: 75-220.

Mendaki (1982a), *Kongres Pendidikan Anak-anak* (Mendaki Congress Papers).

_____(1982b), 'Keluarga Bahagia' (internal document).

_____(n.d.), 'Utamakanlah Belajar di Rumah' (pamphlet).

Mokhtar Abdullah (1968-70), 'The Value of Malay Education in Singapore', *Intisari*, 3(3): 13-19.

Muslim Religious Council of Singapore (MUIS) (1984), *Muslims in Singapore: A Photographic Portrait*, Singapore, MUIS and MPH.

Nagata, Judith (1975), 'Perceptions of Social Inequality in Malaysia', *Contributions to Asian Studies*, 7: 113-36.

_____(1976), 'Kinship and Social Mobility among the Malays', *Man*, N.S., 11: 400-9.

_____(1979), *Malaysian Mosaic*, Vancouver, University of British Columbia Press.

_____(1980-1), 'Religious Ideology and Social Change: The Islamic Revival in Malaysia', *Pacific Affairs*, 53: 405-39.

National University of Singapore Muslim Society and National University of Singapore Malay Society (1982), 'Transformasi Nilai dan Pembangunan dalam Masyarakat Melayu', paper for Mendaki Congress.

Normala Manap (1983), 'Pulau Seking: Social History and an Enthnography', academic exercise in Sociology, National University of Singapore.

O'Brien, Leslie (1982), 'Class, Gender and Ethnicity: The Neglect of an Integrated Framework', *Southeast Asian Journal of Social Science*, 10(2): 1-12.

Pang Eng Fong (1976), 'Growth, Equity and Race', in Riaz Hassan (ed.), *Singapore: Society in Transition*, Kuala Lumpur, Oxford University Press, pp. 326-38.

_____(1983), 'Race, Income Distribution and Development in Malaysia and Singapore', in Linda Y. C. Lim and Peter Gosling (eds.), *The Chinese in Southeast Asia*, Singapore, Maruzen Asia, Vol. 1, pp. 316-35.

Pang Eng Fong and Linda Lim (1976), 'The School System and Social Structure in Singapore', *Commentary*, 4: 43-7.

Pang Keng Fong (1983), 'The Malay Royals of Singapore', academic exercise in Sociology, National University of Singapore.

Parry, Jonathan (1986), 'The Gift, The Indian Gift and the "Indian Gift"', *Man*, N.S., 21: 453 73.

Provencher, Ronald (1971), *Two Malay Worlds: Interaction in Urban and Rural Settings*, Berkeley, University of California, Centre for South and Southeast Asian Studies.

Radcliffe, David James (1970), 'Education and Cultural Change Among Malays 1900-1940', Ph.D. dissertation in Modern History, University of Wisconsin.

Rahim Ishak (1979), 'Teachings of Islam', *Speeches*, 2(9): 46-9.

Ramsay, A. B. (1956), 'Indonesians in Malaya', in *Journal of the Malayan Branch of the Royal Asiatic Society*, 29(1): 119-24.

Riaz Hassan (1977), *Families in Flats: A Study of Low Income Families in Public Housing*, Singapore, Singapore University Press.

Roff, William R. (1967), *The Origins of Malay Nationalism*, New Haven, Yale University Press.

Sahlins, Marshall (1974), *Stone Age Economics*, London, Tavistock.

Saim Noordin (1965), 'Sapintas Lalu...', in *Penyata Tahun 1965*, Singapore, Persatuan Persuratan Pemuda Pemudi Melayu (PPPPM).

Salaff, Janet W. (1981), *Working Daughters of Hong Kong*, Cambridge, Cam-

bridge University Press.
Salaff, Janet W. and Wong, Aline K. (1976), 'Women's Work in Singapore: A Handle for Smaller Family Size', paper presented at Wellesley Conference on Women and Development.
SANA (Singapore Anti Narcotics Association), *Annual Reports*, 1979, 1980, 1982.
Saw Swee Hock (1970), *Singapore Population in Transition*, Philadelphia, University of Pennsylvania Press.
Scott, James C. (1972), 'The Erosion of Patron-Client Bonds and Social Change in Rural Southeast Asia', *Journal of Asian Studies*, 32(1): 5-37.
Senu Abdul Rahman (1971), *Revolusi Mental*, Kuala Lumpur, Penerbitan Utusan Melayu.
Sharom Ahmat and Wong, James (eds.) (1971), *Malay Participation in the National Development of Singapore*, Singapore Community Study Centre and Central Council of Malay Cultural Organizations.
Siebel, Maureen (1961), 'The Changes in the Malaysian Population of Singapore 1819-1959', academic exercise, University of Singapore.
Singapore: A Salute to Singapore, (1984), Singapore, The Times of Singapore.
Smith, W. R. (1975), 'Beyond the Plural Society: Economics and Ethnicity in Middle American Towns', *Ethnology*, 14(3): 225-42.
Steinberg, Stephen (1981), *The Ethnic Myth*, New York, Atheneum.
Stivens, Maila (1981), 'Women, Kinship and Capitalist Development', in Kate Young, Carol Wolkowitz, and Roslyn McCullagh (eds.), *Of Marriage and the Market*, London, CSE Books.
Stoler, Ann (1977), 'Class Structure and Female Autonomy in Rural Java', in Wellesley Editorial Committee, *Women and National Development: The Complexities of Change*, Chicago, University of Chicago Press, pp. 74-89.
Stough, John (1983), 'Chinese and Malay Factory Workers: Desire for Harmony and Experience of Discord', in Linda Y. C. Lim and Peter Gosling (eds.), *The Chinese in Southeast Asia*, Singapore, Maruzen Asia, Vol. 2, pp. 231-55.
Sussman, Marvin B. and Burchinal, Lee (1962), 'Kin Family Network: Unheralded Structure in Current Conceptualizations of Family Functioning', *Journal of Marriage and the Family*, 24: 231-40.
Swift, Michael J. (1963), 'Men and Women in Malay Society', in Barbara E. Ward (ed.), *Women in the New Asia: The Changing Social Roles of Men and Women in South and Southeast Asia*, Paris, UNESCO, pp. 268-86.
_____(1964), 'Capital, Saving and Credit in a Malay Peasant Economy', in Raymond Firth and B. S. Yamey (eds.), *Capital, Saving and Credit in Peasant Societies*, London, George Allen & Unwin, pp. 133-56.
_____(1965), *Malay Peasant Society in Jelebu*, LSE Monographs on Social Anthropology 29, London, Athlone Press.
_____(1967), 'Economic Concentration and Malay Peasant Society', in Maurice Freedman (ed.), *Social Organization, Essays Presented to Raymond Firth*, London, Frank Cass, pp. 241-69.
Syed Husin Ali (1964), *Social Stratification in Kampong Began*, Malaysian Branch of the Royal Asiatic Society Monograph 1, Singapore, Malaysian Printers.
Tan Jin Lee, Winifred (n.d.), 'Chinese Kinship under Change in Singapore', academic exercise in Sociology, University of Singapore.
Tan Keng Yam, Tony (1981), 'Speech by Dr. Tony Tan Keng Yam, Minister for Trade and Industry, at MUIS Bursaries Presentation Ceremony held at D.B.S. Auditorium, Wednesday 2.9.1981', *Warita*, 18: 10-11.

Thadani, Veena N. (1978), 'The Logic of Sentiment: The Family and Social Change', working paper, The Population Council, New York.
Tham Seong Chee (1983), *Malays and Modernization*, rev. ed., Singapore, Singapore University Press.
Tunku Shamsul Bahrin (1967a), 'The Pattern of Indonesian Migration and Settlement in Malaya', *Asian Studies*, 5(2): 233-57.
_____ (1967b), 'The Growth and Distribution of the Indonesian Population in Malaya', *Bijdragen*, 123: 267-86.
_____ (1970), 'The Indonesian Immigrants and the Malays of West Malaysia', *Geographica*, 6: 1-12.
Turnbull, C. M. (1977), *A History of Singapore 1819-1975*, Kuala Lumpur, Oxford University Press.
Vrendenbregt, Jacob (1964), 'Bawean Migration: Some Preliminary Notes', *Bijdragen*, 120(1): 109-37.
Wahidah Jalil (1981), 'A Study of Muslim Divorce in Singapore', academic exercise in Sociology, National University of Singapore.
Wallerstein, Immanuel, Martin, William G., and Dickenson, Tony (1982), 'Household Structures and Production Processes: Preliminary Thesis and Findings', *Review*, 5(3): 437-58.
Wallman, Sandra (1978), 'The Boundaries of "Race": Processes of Ethnicity in England', *Man*, 13(2): 200-17.
Ward, Barbara E. (1963), 'Men, Women and Change: An Essay in Understanding Social Roles in South and Southeast Asia', in Barbara Ward (ed.), *Women in the New Asia: The Changing Social Roles of Men and Women in South and Southeast Asia*, Paris, UNESCO, pp. 25-99.
Weber, Max (1930), *The Protestant Ethnic and the Spirit of Capitalism*, trans. Talcott Parsons, London, George Allen & Unwin.
Wee, Ann (1963), 'Chinese Women of Singapore: Their Present Status in the Family and Marriage', in Barbara Ward (ed.), *Women in the New Asia: The Changing Social Roles of Men and Women in South and Southeast Asia*, Paris, UNESCO, pp. 376-409.
Wee, Vivienne (1976), ' "Buddhism" in Singapore', in Riaz Hassan (ed.), *Singapore: Society in Transition*, Kuala Lumpur, Oxford University Press, pp. 155-88.
_____ (1984), 'Material Dependence and Symbolic Independence: Constructions of Malay Ethnicity in Island Riau, Indonesia', paper presented at the Conference on Ethnic Diversity and the Control of Natural Resources in Southeast Asia, University of Michigan.
_____ (n.d.), 'Religious Patterns in Singapore: A Panoramic View', paper presented at the Symposium on Social Anthropology in Complex Societies: the Case of Singapore, XIth International Congress of Anthropological and Ethnological Sciences.
White, Benjamin (1976), 'The Economic Importance of Children in a Javanese Village', in David Banks (ed.), *Changing Identities in Modern Southeast Asia*, Paris and The Hague, Mouton, pp. 269-90.
White, Benjamin and Hastuti, Endang Lestari (1980), 'Different and Unequal: Male and Female Influence in Household and Community Affairs in Two West Javanese Villages', working paper 06, Agro-Economic Survey, Rural Dynamics Study and Centre for Rural Sociological Research, Bogor Agricultural University, Bogor.
Whitehead, Ann (1981), ' "I'm Hungry Mum": The Politics of Domestic

Budgeting', in Kate Young, Carol Wolkowitz, and Rosyln McCullagh (eds.), *Of Marriage and the Market: Women's Subordination in International Perspective*, London, CSE Books, pp. 88-111.

Wilder, William (1970), 'Socialization and Social Structure in a Malay Village', in Philip Mayer (ed.), *Socialization: The Approach from Social Anthropology*, London, Tavistock, pp. 215-68.

_____ (1982a), *Communication, Social Structure and Development in Rural Malaysia*, LSE Monographs on Social Anthropology 56, London, Athlone Press.

_____ (1982b), 'Psychosocial Dimensions of Ethnicity', *Southeast Asian Journal of Social Science*, 10(1): 103-15.

Willis, Paul (1977), *Learning to Labour*, London, Saxon House, reprint 1980, Westmead, Gover.

Winstedt, Sir Richard (1950), *Malay Proverbs*, London, John Murray.

Wong, Charlotte Hock Soon (1974), 'The "Little Businessmen"', Masters dissertation in Sociology, University of Singapore.

Wong, Diana (1983), 'The Social Organization of Peasant Reproduction: A Village in Kedah', Ph.D. dissertation in Sociology, Bielefeld.

Yang, Martin (1967), 'The Family as a Primary Economic Group (China)', in G. Dalton (ed.), *Tribal and Peasant Economics*, New York, Natural History Press, pp. 333-46.

Yang Hon Loon (1973), 'The Practice of Nepotism: A Study of Sixty Chinese Commercial Firms in Singapore', academic exercise in Sociology, University of Singapore.

Yang Razalie Kassim (1979), 'Education and the Malays of Singapore 1959-1979', academic exercise in Political Science, National University of Singapore.

Yee, Janet (1959), 'Disputes Among Singapore Chinese Families', research paper in Social Studies, University of Malaya (Singapore).

Young, Kate, Wolkowitz, Carol, and McCullagh, Roslyn (eds.) (1981), *Of Marriage and the Market: Women's Subordination in International Perspective*, London, CSE Books.

Young Mei Ling and Kamal Salih (1984), 'The Malay Family: Structural Change and Transformation', paper presented at the Research Seminar on Third World Urbanization and the Household Economy, Universiti Sains Malaysia, Penang.

Zahoor Ahmad bin Haji Faizal Hussain (1969), 'Policies and Politics in Malay Education in Singapore 1951-1965, with Special Reference to the Development of the Secondary School System', Master of Education dissertation, Singapore.

Singapore: Official Documents, Censuses, and Surveys

Official Documents

State of Singapore (1960), *Report of the Commission of Inquiry into the System of Contract Labour in Singapore*.

Singapore: Administration of Muslim Law Act 1970 Edition, Republic of Singapore, Government Printer.

Singapore: Syariah Court Appeals 1 & 2, 1978.

Parliamentary Debates, various years.

Censuses

A Report on the 1931 Census of British Malaya, London, Crown Agents. (Abbreviated in references as 'Census 1931'.).
Report on the Census of Population 1957, Singapore, Government Printer. (Abbreviated in references as 'Census 1957'.)
Census of Population 1970, Singapore, Singapore, Government Printing Office. (Abbreviated in references as 'Census 1970'.)
Report on the Census of Population 1980, Parts 1-9, Singapore, Government Printing Office. (Abbreviated in references as 'Census 1980'.)

Surveys and Statistics

Report on the Household Expenditure Survey (1972-73), Department of Statistics, 1974.
Report on the Labour Force Survey, Ministry of Labour, 1975, 1978, 1980.
Singapore Yearbook of Labour Statistics, Research and Statistics, Ministry of Labour, 1970, 1980.
Report on the Census of Wholesale and Retail Trades, Restaurants and Hotels, Department of Statistics, 1973.
Report on the Survey of Wholesale and Retail Trades, Restaurants and Hotels, Department of Statistics, 1977, 1981.
Economic and Social Statistics of Singapore, Department of Statistics, 1960-82.
Key Educational Statistics, Ministry of Education, 1981, 1982, 1983.

Newspapers and Periodicals

Far Eastern Economic Review (abbreviated in text as *FEER*).

6.8.76	Sterilization and Literary
1.8.80	Singapore 1980 (Focus)
31.7.81	Singapore 1981: MOE and the Real Cost of Academic Freedom
2.8.83	The Price of Education
8.8.83	Designer Genes
15.9.83	Patient, Heal Thyself
6.10.83	Big Fish, Small Pond
20.10.83	The Minority's Dilemma
1.12.83	Stratified Society
8.12.83	No Sacred Cows
16.2.84	The Case for the Malays
1.3.84	The Dangers of Eugenics
26.4.84	More on Minorities
17.5.84	A May Day Signal
17.5.84	Pride and Prejudice
21.6.84	Some are More Equal
28.6.84	Joining the Mainstream
26.7.84	The Malay Solution
16.8.84	The Malaysian Way

The Mirror

1.2.83 Malay Singaporeans—Why the Socioeconomic Gap 19(3), 1-4.

New Nation

5.9.81   Education Boost for Malays

Singapore Herald

6.8.70   Two Lies that Keep Us Apart
7.8.70   Vicious Circle that's still Unbroken
10.8.70   Ideas for Aid that Fit Social Facts
11.8.70   Stroke Two of a Two-stroke Mechanism

Sunday Monitor

21.3.83   The Super Breed?
30.9.84   Malays are Losing Proficiency in the Use of their Language: Zulkifli

Straits Times or Sunday Times (abbreviated in text as *ST*)

4.9.78   Rise of the New Malay
8.9.78   Help Malays to Help Themselves
15.9.78   Malay Dilemma of Modernization
19.3.79   Existing Problems in the Education System
28.3.79   Dr. Goh on Streaming
29.7.79   Registration of Pre-primary and Primary One Classes
21.8.79   Some Push Stream, Others Proximity
1.9.79   The Power of Education and Dr. Goh: Help These Students
11.9.79   The Restructured Education System: An Interview
22.8.80   Chok Tong Urges More Malays to Join Varsity
10.9.80   Malays' Changed View of Education
2.12.80   Day a Kampong Was Filmed
11.2.81   Streaming and Language: The Untested Claims
19.2.81   Dialect Home Kids Did Poorly in PSLE
6.3.81   Streaming, Language and Mobility–The Facts and Figures
23.3.81   Rahim on Why No Muslim Aged are in Homes
4.4.81   No Dispute about Objectives, but Are Methods Right
4.4.81   A Case for Subject Streaming in Our Secondary Schools
29.4.81   Streaming in fact Gives a Better Chance to Weak Pupils from Poorer Homes
22.8.81   Compulsory School for All between 6 to 18?
23.8.81   Chok Tong: Get as Many as Possible into Secondary Schools
17.1.82   Best Average Bet Yet
20.1.82   Educating Our Children for a Better Future
20.1.82   To Upgrade Skills of Malay Youth Devan Calls for a United Effort
4.2.82   Publish Synopsis of Syllabus for Public Debate: Dr Goh
9.3.82   Leaders to Draw Plan for Helping Muslim Pupils
18.4.82   Let Studies Come First Says Mattar
16.5.82   School the Main Battlefront for Tackling Problems
23.5.82   Higher Education to Keep up with Others
29.5.82   Students and Parents Must Make the Effort
29.5.82   Malay Pupils Caught in a Vicious Circle
4.7.82   Parents and Principals Are More Receptive to Streaming

22.7.82	Muslims Urged to Change Attitudes on Education
27.7.82	What Price Our High Achievers
9.8.82	Malays Have Made Great Strides for the Better, says Ow
17.8.82	Role of the Family in Child's Education
18.8.82	The Lessons Are Clear–Let's Learn from Them: PM
16.9.82	Call to Take More Interest in Malays
7.11.82	Look into Ways to Solve Problems, Malays Told
2.12.82	Minister on How to Give the Slow Learners a Better Deal
26.4.83	When Troubles Come after the Wedding
5.6.83	Towards a Smart Society
2.8.83	The Price of Education
25.8.83	Way to Keep Singapore Going
28.8.83	Streaming Has Succeeded in Cutting Number of Premature School Leavers
6.11.83	It Works
22.1.84	Study Shows Importance of Parents' Education to Children
24.1.84	School Entry Rules Changed
13.3.84	Dr. Goh: Go Look for the Classless Society
3.6.84	All for the Good of the Children of Our Children
19.7.84	Parental Support Goes a Long Way
6.8.84	Why the Old and Young Go to Syariah Court
15.8.84	Thumbs up for Kampong Project
20.9.84	Students with Grad Mums Perform Better
29.9.84	Panel to Tackle Malay Problems
11.10.84	Grad Mums are Best

Index

ABDULLAH BIN M. BAGINDA, 95
Abercrombie, Nicholas and Turner, Bryan S., 181
Administration of Muslim Law Act, 18; *harta syarikat*, 30, 35; inheritance, 70
Adoption, 14-15; children's obligations, 44; strategy for old age care, 64
Adulthood, transition on marriage, 55, 59
Ahmad Ibrahim, 30
Ahmad Mattar, 172
Alatas, Syed Hussein, 137-8, 162, 168
Alexander, J. and Alexander, P., 131
Aljunied, S. Zahra, 103, 109
Ancestors: Malay, 69; Chinese, 162
Apprenticeship, 139
Arab, 98, 168
Assistance: between Malay kin, 130; between Chinese kin, 151, 159-60
Authority: in Chinese household, 153-5; in Malay parent-child relations, 56-9; through arranged marriage, 58-9

BAILEY, CONNOR, 5-6, 65, 131-2, 147
Bangkit: in family business, 85; between siblings, 82
Banks, David J., 7, 8, 30, 33, 36-7, 42, 65, 147
Bawean, migrants from, 93, 95
Bedlington, Stanley S., 98, 170-2
Belanja kahwin, 38, 59
Benjamin, Geoffrey, 92, 98
Betts, Russell Henry, 94, 98, 108, 137, 170
Bilateral kinship, 185
Block committee, 128-9, 135
Blood relations, 7
Bonded labourers, 94
Bourdieu, Pierre, xvi, 9-11, 168
British services: cause Malay migration, 94; recruit Malays, 108
Brook, Eve and Finn, Dann, 182n
Budget, *see* Domestic budget
Bureaucratization, 123-4
Business, *see* Entrepreneurship

CALDWELL, J. C., 63, 73-4, 185
Capitalism, commodification, 10; *see also* Commodification
Carstens, Sharon, 151
Central Provident Fund: on divorce, 26; effect on fertility, 64, 123; reduces disposable income, 113
Chang Chen Tung *et al.*, 14, 49, 106, 152
Chang Soo, 157, 159
Chayanov, A. V., 6
Cheah Hock Beng, 112, 181
Chee Heng Leng and Chan Chee Khoon, 121n
Chen, Peter S. J. *et al.*, 150
Cheng Siok Hwa, 104
Child transfers, 14; *see also* Adoption
Child-care, 15; as business, 143; costs, 45; grandmother's gift, 45; pleasures, 43; social class, 125
Children: focus of household, 123; responsibility for family welfare, 48; value, 41, 72-4; Chinese, 155; Malay and Chinese, 156
Chinese: associations, 97, 151, 160-1; customary law, 150; household labour, 108; household maintenance, 39; interaction with Malays, 136; labour contracting, 97; migration patterns, 96-7; perception of Malays, 110-11; sub-groups, 136
Chinese language: in business, 108; in education, 118; relation to cultural values, 179
Chua Beng Huat, 151, 162n, 182
Chung, Betty Jamie *et al.*, 153-5
Clammer, J., 180-2
Commodification: definition, 10; and the gift, 11; in husband-wife relationship, 25; in parent-child relationship, 45-7
Communications, within household, 17
Communitarianism, 130-3
Confucianism, 179
Consumer spending, 171

Consumption rights, 15
Corporate family economy, 8; Malay, 41, 75; Chinese, 150
Cultural heritage, xv, xvi; and historical change, 4
Cultural involution, 92, 134-6
Cultural values, as explanations for Malay position, 166-77; Chinese, 179-80
Culture of poverty, 182n
Current reproduction, 5, 40

DARAH, 7
Data: collection, vii-viii; types, xvi-xvii; ethnographic, xvi, 92, 99, 186; quantitative, xvii, 92, 99, 186
Daughter: Chinese, 151-8
Daughter: Malay, care for elderly, 62-3; contribution to household, 49-53; financial independence, 59; relationship to mother, 50-1; unpaid domestic services, 49
Debts, relation to gifts, 9
Devan Nair, 174
Dewey, Alice G., 6, 86, 147, 162n
Discrimination: by Chinese employers, 109-11; by government, 108-9, 121n; in business, 140, 161; in education, 119; legitimation of inequality, 179
Divorce rate: Malay, 14; measurement, 34; decline, 35
Divorce: Chinese, 150
Divorce: Malay, cultural changes, 35; division of property, 25; effect on householding relations, 27; failure to provide maintenance, 18, 138; infertility, 36; personal savings, 26; primary data source, viii; relation to adoption, 14; residence patterns, 37
Divorced men, and household, 16
Divorced women: economic vulnerability, 35; relations with children, 66; migration, 96
Djamour, Judith, 8, 14-15, 25, 28, 30, 35-6, 41, 44, 57, 69, 71, 86, 94
Domestic budget, 20, 103; in divorce, 37; use of gift idiom, 44
Domestic services: as gifts, 23-5; commercial value, 10, 23; commodification, 44-7; provided by unmarried daughters, 49; undervalued commodity, 24-7
Dominant ideology thesis, 181
Drug abuse, 54, 183n; national service recruitment, 109
Dzulfiqhar Mohd. M., 183n

ECONOMIC CONDITIONS, xv, 4
Economic position of Malays and Chinese: pre-1959, 100, 166; post-1959, 102-3, 166
Education: Chinese culture, 134; Chinese family, 119-20, 157-8; expenses, 121; fertility, 106; Malay family, 119-20, 157-8; Malay problem, 169-70; Malay values, 173-7; meritocracy, 114-16; old age insecurity, 80; parental investment, 79-82; policies, 121n; social class, 81, 118, 120, 121n, 125
Educational status, Malays and Chinese: pre-1959, 101-2; post-1959, 117-20
Egalitarian ethic, 129-33
Elementary family, Malay: co-residence, 13-14; business, 160; intensified relations, 123-4; marital stability, 36
Elementary family, Chinese, 158-9; in business, 159-60
Élite: Chinese, 101, 166, 180; Malay 101, 166, 168, 173, 178-9
Entrepreneurship: Chinese family, 158-62; direct selling, 144-6; economic returns, 107; employment opportunities, 107-8; ethnic marker, 137, 166-8; in Java and Malaysia, 14; Malay and Chinese participation 1957-80, 106; Malay family labour and capital, 84-8; Malay image, 137, 166-8; Malay motivation to acquire wealth, 83; Malay social relations, 141-8
Ethnic groups, and data presentation, 99, 174
Ethnic interaction, 136
Ethnic sentiment, 97, 98, 134-6; in business, 137-48
Ethnicity, as ideology, 181-2
Eugenics, 114, 183n
Expanded reproduction, 5, 75
Extended reproduction, 5, 40
Eysenck, 121n

FAMILY LABOUR AND CAPITAL: in Malay business enterprise, 84-8
Family nucleus, definition in Singapore census, 13
Farmers: 1931, 93; 1957-80, 102; Malay image, 170-3
Fatalism, 171
Father-daughter relationship, 52-3
Father-son relationship, 53-4
Feminist analysis, 5; *see also* Women's studies
Fertility rate, 14; decline, 43, 64; Malay and Chinese 1957-76, 105
Fertility: care of elderly, 63; relation to class, 123
Filial piety, in Chinese family, 150, 154-9

Fishermen: 1931, 93; 1957-80, 102; Malay image, 170
Food business: Malay 139, 141; Chinese, 159
Freedman, Maurice, 97, 149-53, 159, 161-2
Funeral expenses, 128

GANS, HERBERT, 182n
Geertz, Clifford, 6, 83, 130-1, 147
Geertz, Hildred, 16, 31-5, 39, 44, 58
Gellner, Ernest, 183
Gender, 5; *see also* Household; Women; Women's studies
Generalized reciprocity, 9
Gerakan Obor, 169
Geylang Serai, 95
Giddens, Anthony, xvi, 10, 167
Gift, 8-9; commercial values, 11; individualism, 8; in husband-wife relations, 23; in parent-child relations, 41-4; Mauss, 7; *see also* Inheritance
Goh Chok Tong, 173-4
Goh Keng Swee, 100-1, 112, 180
Gopinathan, S., 102, 121n, 183n
Gordon, Shirle, 34, 170
Gordon, Shirle and Gunn, G. T., 118
Gotong-royong, 132
Government, *see* Singapore government
Graduate women, government policies, 115
Grandmother, role in child-care, 44, 46
Great traditions, 97

HAJ: use of wealth, 79; and migration to Singapore, 94
Hanna, W. A., 166, 183n
Hardjono, Joan, 55, 131-2
Harevin, Tamara, 12n
Harris, Olivia, 6
Harris, Olivia and Young, Kate, 12n
Hart, Gillian, 55, 131
Harta syarikat, harta sepencarian, 29-31; in Malaysia and Java, 29-31
Hartman, Heidi, 5, 12n
Harun Aminurrashid, 166
Hegemony, 181-2
Historical analysis, xv-xvii, 167, 184
Household Research Working Group, 3-5, 12n
Household: Chinese business, 158-62; Malay business, 83-8; definition in Singapore census, 13; educational performance, 79-82, 119-20; historical change, 3, 4; husband-wife as core, 39; income pooling, 5; membership, 13-17; size, 104-6; social class variation, 123-4; sociological debates, 3; unit of production and consumption, 5

Householding, 4; divergent interests, 5
Husband: adjusting to marriage, 38; borrowing from wife, 19; duty to provide maintenance, 18; gifts to wife, 21-2; secret saving, 22
Husband-wife, partnership, 75, 87, 40; *see also* Harta syarikat
Husbands, ideology, definition, 167

INCOME, MALAYS AND CHINESE: pre-1959, 100, 101; post-1959, 102, 103
Independence, impact on Malays, 98
Indian Muslim, 168
Indirect generation reciprocity, 69
Individual, and salvation, 76
Individualism, 6; business, 85, 162n; divorce, 34; family relationships, 33; household, 184-5; kinship sentiment, 9; limits accumulation, 76; migration, 96; property ownership, 8, 70-1; relations in community, 127, 129-33; success in education, 80; the gift, 9; value of children, 73
Infertility, and divorce, 36
Informal enterprise, 107, 139, 141; women's work, 24
Inheritance, 7; as a gift, 42, 70-2, 78; land versus education, 78; parental role undermined, 79; postponed, 65; relation to salvation 77; relationship with deceased, 69-72
Islam: business ethics, 140; care for elderly, 65-6; community relations, 127-9; favours entrepreneurship, 83; ingrained in Malay culture, 7; investment, 85; Malay backwardness, 170, 172-3; Malay problem, 183n; Malay progress, 176-7; modernizing potential, 183n; parent-child relations, 42; property division on divorce, 31; protestant ethic, 83; salvation, 67-71; Malay identity, 97
Ismail Kassim, 121

JAPANESE CONSCRIPTS, 95
Java: children's obligations, 44; comparative data, xvi; divorce rate, 34; entrepreneurship, 147; *harta syarikat*, 29-31; household membership, 14; husband-wife co-operation, 32; labour contribution of boys, 55; labour contribution of girls, 49; migrants from, 7, 93-5; Muslim merchants, 83; old-age support, 65; prayers for the deceased, 68-9; village relations, 131-2
Jay, Robert, 14, 31, 44, 55, 65, 68-9, 132
Jenkins, Richard, 12n
Jodoh, 33, 36
Johore, settled by Javanese, 95

INDEX

KAMAL SALIH *et al.*, 5, 12n
Kampong: ideal life style, 129; urban orientation, 96
Kampong Java, 94, 138
Kasihan: definition, 11; expressed by children, 48, 58; expressed by parents, 57; provision of child-care, 45-6
Kenduri Arwah, *Kenduri Ruahan*, 69, 71-2
Kesayangan, *see* Kinship sentiment
Kin relations: Chinese, 151; in Chinese business, 159-60; Malay, in Singapore, 123, 130; Java, 131-2; Malaysia, 131-2; in Malay business, 84-9
Kinship ideas, 6-11; shared poverty, 130-1
Kinship sentiment: basis for householding, 184-5; definition, 7; family business, 86; husband-wife relations, 23; importance to propertyless wage earners, 65; individual will, 8; parent-child obligations, 41; relations with deceased, 71; value of children, 72-4; valued by elderly, 64, 66-7
Koentjaraningrat, R. M., 132

LABOUR FORCE PARTICIPATION: women, 49, 105; Chinese women, 152; married Malay women, 23-4, 27; unmarried Malay women, 49, 105
Lai, Alice, 158, 171
Language competence: constraint to business, 140; education and social class, 115-16; neighbourhood interaction, 136
Lau Hong Thye, 139
Lee Kuan Yew, 114-15, 118, 121n, 150, 173-4
Legitimization of inequalities, 167, 178-82; open education, 113-15, 179
Leong Choon Cheong, 39, 125, 158, 171, 179-81
Leong Pui Ling, 162n
Lewis, Oscar, 182n
Liebow, E., 39
Life expectancy, 14
Lim, Linda Y. C., 161
Lim Guek Poh, 154
Lim Kim Huay, 139
Little traditions, 97

MA'AROF SALLEH, 183n
Macfarlane, Alan, 6, 36, 41, 73, 74n, 185
McGee, T. G., 6, 12n
McGlothlin, W. H., 54
McKinley, Robert H., 14, 15, 42, 44, 86, 132
McLellan, Susan, 49, 52
Madge, Charles, 161
Mahathir bin Mohamed, 121n, 161, 169

Mak Lau Fong, 151
Malay Dilemma, 169
Malay education: colonial policy, 101; in Malay language, 118-19
Malay language, Singapore Malay identity, 97
Malay nationalism, 98
Malay Youth Literary Association, 169
Malaysia: care of elderly, 65-6; comparative data, xvi; divorce rate, 34; entrepreneurship, 147; *harta syarikat*, 29-31; household membership, 14; labour contribution of boys, 55-6; labour contribution of girls, 49, 52; migrants from, 7, 93-5; residence patterns and divorce, 36-7; urban kin relations, 132; village relations, 131-2
Malaysian government, favours Malays, 111
Mansor Haji Sukeimi, 121n, 183n
Marriage counselling, 18, 36, 183n; *see also* Syariah Court
Marriage: arranged, 35, 58-9; statistical and normative instability, 34; property ownership, 8; rising age at, Chinese, 153; rising age at, Malay girls, 48-9, 51; Malay boys, 54; *see also* Divorce
Marx, Karl, xv
Matrifocal family, 39
Mauss, Marcel, 7, 9
Maxwell, Hon. W. E., 29
Mendaki, 175-7, 183n
Meritocracy, 169; in education, 114-16, 121; social class, 178-9
Migration to Singapore, 93-6; to Malaysia, 111
Mission schools: pre-1959, 101-2; and social class, 116, 135
Mokhtar Abdullah, 169
Mother-daughter relationship, 50-1
Mother-son relationship, 53
Multinationals, Malay women, 104
Muslim reformers, 67, 69, 76, 168, 173
Muslim Religious Council of Singapore, 173

NAFKAH: definition, 18; reasons for prominence in Singapore, 31-2
Nagata, Judith, 101, 132, 146-7, 161, 180, 183
National University of Singapore Muslim and Malay Societies, 172
National Wage Council, 113
Neighbours: Chinese, 134-6; Malays, 135; rehousing, 123; social circle, 124-9; social class, 124-9; weddings and funerals, 126-9
Normala Manap, 124

OCCUPATIONAL DISTRIBUTION: Malays and Chinese 1957-80, 102
Old-age: Chinese, 150; Malay, fertility strategy, 63; independence, 60; insecurity, 60-6; insecurity and education of children, 80; *kasihan*, 60; reliance on daughters, 62; savings, 62-3
Orang baik, 126-9
Origins of Malay community, 93-6
Orthodoxy, definition, 168

PAHALA, 67-9
Pang Eng Fong, 183n
Parent-child relationship, social class variation, 47, 56
Parenthood: defined, 42; flow of wealth and generosity to children, 42, 46; social class variation, 47, 56; undermined, 57, 59
Parry, Jonathan, 11
Partnership: husband-wife, 20; contractual, 29
Party system business, 144-6
Peasant household, 6
Peasant society, 131
Peranakan, 101
PERDAUS, 183n
Pilgrim broker, 138; migration to Singapore, 94-5
Poverty: in divorce, 39; increased post-1959, 112; rural village, 131; standards, 100
PPPPM, 169
Prayers for the deceased, 67-72
Property division, on death, 70; on divorce, 25-33
Property ownership: individual, 8; old age support, 65-7; prayers for deceased, 71
Protestant ethic, 83
Provision shops, 146
Publications, 138

RAHIM ISHAK, 173
Reciprocity: among villagers, 132; business relations, 145-6; exchange, 11; indirect, 69
Rehousing: Chinese families, 151; effect on household form, 123; disposable income, 113; social class, 128
Reification, 167
Religious education: daughters, 59; neighbourhood activity, 128; old age care, 66; parental duty, 67; prayers for the deceased, 67-8; sons, 56
Reproduction: current, 40; expanded, 75; extended, 40; types defined, 5
Research methods, vii

Residence pattern, 13-15; cause of divorce, 37
Restu, 177
Revolusi Mental, 169
Rezeki, 171
Riau Islands, migrants from, 93
Riaz Hassan, 150, 155, 158
Roff, William R., 93, 95, 98, 101, 166
Royalty, 93
Rural employment in Singapore: 1931, 93; 1957-80, 102
Rural image, 93-4, 138, 166, 170-1, 173, 183n

SAHLINS, MARSHALL, 6, 43
Saim Noordin, 169
Salaff, Janet W., 39, 152, 154
Salaff, Janet W. and Wong, Aline K., 151-2, 154
Salvation: Chinese, 162; Malay, relation to wealth, 76-8
Sample households, vii, viii
Savings, security in old age, 62
Schools, good and bad, 116
Scott, James C., 131
Sea-nomads, 93
Senu Abdul Rahman, 169
Separation from Malaysia, 98, 111
Sha'ari Tadin, 182n
Shared poverty, 130-3, 146-7
Sharom Ahmat and Wong, James, 118, 170
Siblings: Chinese, 157; Malay, rivalry, 81-3; co-operation in business, 84; separate economies, 75
Siebel, Maureen, 93, 94
Singapore Constitution, and Malay status, 111
Singapore government: Chinese cultural values, 179; defence services, 108-9; economic programme, 113; impact on household, 124; Malay problem, 174; policies, 182
Singapore Herald, 172
Skinner, 161
Smith, W. R., 182n
Social class: Chinese family, 155-6; cultural concepts, 124-36; defined, 99-100; education, 81, 102, 114-16, 158, 174-7, 120, 121n; educational streaming, 116; ethnicity, 112, 134-6; household form, 123-4; institutionalized, 122; language competence, 115; legitimation, 178-82; neighbourhood interaction, 178; parent-child relations, 47, 56; relation to fertility, 106, 123; social circle, 124-9
Social relations: kin and neighbours, 131-3; entrepreneurship, 141-8

Sombong, 126-9
Sons: Chinese, 151-8; Malay, contribution to household budget, 53-6; care for elderly, 62-3; presence in home, 16
Statistical data, xvii
Steinberg, Stephen, 182n
Sterilization, government policies, 115
Stivens, Maila, 12n
Stoler, Ann, 131
Stough, John, 121
Streaming, relation to class, 116-18
Sub-contracting, 140
Sussman, Marvin B. and Burchinal, Lee, 12n
Swift, Michael J., 28, 30, 131, 133, 162, 162n
Syariah Court, and *harta syarikat*, 30
Syariah Court: property settlement, 25, 70; wife's right to *muta'ah*, 27; *see also* Marriage counselling
Symbolic systems, 167
Symbolic violence, 9

TAN, TONY, 173
Tan, Winifred, 151-2, 162n
Thadani, Veena N., 3, 12n, 73
Tham Seong Chee, 137
The Mirror, 173
Theoretical framework, xv
Trading community, Malay and Javanese, 138
Tunku Shamsul Bahrin, 94
Turnbull, C. M., 95, 98, 102

UMNO, 169
Unskilled labour, displaced post-1959, 112-13
Urban: settlement, 95-6, 166; employment, 95, 100

VALUE OF CHILDREN, 41, 72-4; Chinese, 155; Malay and Chinese, 156
Vrendenbregt, Jacob, 95

WAHIDAH JALIL, 14, 34, 49, 150
Wallerstein, Immanuel *et al*., 12n
Ward, Barbara E., 162n
Wealth: accumulation, 76-8, 83, 184; Chinese family, 161-2
Weber, Max, 83
Wedding: business aspects, 87-8; costs, 171
Wee, Ann, 151, 154, 162n
Wee, Vivienne, 171
White, Benjamin, 49, 55
White, Benjamin and Hastuti, Endang Lestari, 34
Whitehead, Ann, 5, 12n
Widows, 130
Wife: Chinese, 152; Malay, assistance to husband, 28; contributions as a gift, 19; gifts from husband, 22, 28; personal property, 21; secret savings, 22; use of income, 18; *see also* Domestic services
Wilder, William, 14, 29, 34, 56, 58, 147, 162n
Willis, Paul, 4, 12n, 121
Women, Chinese, 152-3; Malay, domestic services, 10; employment, 23-4, 27, 49, 104-6; *see also* Daughter; Wife
Women's Studies, 5, 24
Wong, Charlotte Hock Soon, 159
Wong, Diana, 6, 12n, 65, 131-2

YANG, MARTIN, 151-2
Yang Hon Loon, 159
Yang Razalie Kasim, 172, 183n
Yee, Janet, 155
Young Mei Leing and Kamal Salih, 5

ZAHOOR AHMAD, 118
Zainal Abidin bin Ahmad, 175